I0024997

Edward John Tilt

The change of life in health and disease

A practical treatise on the nervous and other affections incidental to women at the

decline of life

Edward John Tilt

The change of life in health and disease
A practical treatise on the nervous and other affections incidental to women at the decline of life

ISBN/EAN: 9783337118440

Printed in Europe, USA, Canada, Australia, Japan

Cover: Foto ©berggeist007 / pixelio.de

More available books at **www.hansebooks.com**

THE CHANGE OF LIFE

IN

HEALTH AND DISEASE.

A PRACTICAL TREATISE

ON

THE NERVOUS AND OTHER AFFECTIONS

INCIDENTAL TO

Women at the Decline of Life.

BY

EDWARD JOHN TILT, M.D.

SENIOR PHYSICIAN TO THE FARRINGDON GENERAL DISPENSARY AND LYING-IN CHARITY,
VICE-PRESIDENT OF THE OBSTETRICAL SOCIETY OF LONDON,
FELLOW OF THE ROYAL MEDICAL AND CHIRURGICAL SOCIETY,
AND OF SEVERAL FOREIGN SOCIETIES.

Third Edition.

LONDON:

JOHN CHURCHILL AND SONS,
NEW BURLINGTON STREET.
MDCCCLXX.

[The right of Translation is reserved.]

LONDON:

SAVILL, EDWARDS AND CO., PRINTERS, CHANDOS STREET,
COVENT GARDEN.

PREFACE.

THERE are many milestones on the road that leads from the cradle to the grave—seven, fourteen, and twenty-one are clearly written on the first; forty-two, forty-nine, and sixty-three are distinctly visible, though less deeply cut, on those that mark the last part of the journey. These dates point to periods of the human life characterized by very important changes, many of which give a peculiar aspect to the physiology of the human being, and impart a family likeness to the diseases of epochs justly deemed critical. No division of labour is so praiseworthy as the study of these great periods of human life; and if many of those who fritter away their meritorious energy in writing monographs would devote it to thoroughly investigate the physiology and diseases of the great periods on which nature has put a special stamp, they would very much increase their fame and greatly contribute to the progress of medicine.

When the last census was taken in 1861, 1,177,535 of our fellow-countrywomen of the age of forty-five and under fifty-five were living in Great Britain and Ireland, and I can scarcely exaggerate the importance of a work which tells their history, records the probabilities and the inevitabilities of their future, and investigates the many diseases by which it may be chequered. This volume professes to do so, and it is founded on the tabulated estimates of the symptoms and of the diseases of five hundred women who were at the change of life, or who had passed the ménopause.

In this new edition I have kept to the original plan, giving a paramount importance to physiology, without which, the diseases of any critical period cannot possibly be understood. Although I have retained the statistical information on which the work was based, I have drawn largely on my experience among the upper classes of society during the last thirteen years, so as to rectify any pathological one-sidedness that may have arisen from the statistics having been principally derived from dispensary practice; and I trust I may be permitted to state that, partly because the importance of the subject was recognised by the profession, but chiefly because my work was the only one on the subject in the English language, I have been consulted by a very considerable number of women suffering from diseases of the change of life, since the publication of the second edition thirteen years ago.

To review the labour of rewriting this book is to comment on some of the events of the last few years; thus I have gathered valuable materials for my physiological study of the change of life from the contributions to the International Medical Congress that was held in Paris.

The chapter on the Diseases of the Reproductive Organs has been greatly increased, for I have lost no opportunity of taking careful note of whatever might illustrate this part of the subject, and the reader will there find information he will seek for in vain elsewhere; at least, Dr. Barnes said so of a sketch of the chapter that I read before the Newcastle Meeting of the British Medical Association in 1870.

A reviewer of the second edition, in the *Archives Générales de Médecine*, stated that, although the chapter on diseases of the brain extended to a great length, it was not long enough to do justice to the subject. This is an answer to those who may complain that the chapter is too long.

In the preface of the second edition, I mentioned that "The present century has witnessed magnificent discoveries in the pathology of the brain and of the spinal cord; but it will be obvious to those conversant with medical literature that the pathology of the ganglionic nervous system has received comparatively little attention, neither can it be much advanced until experimental physiologists have accurately

investigated many points connected with the physiology of
the ganglia and their nerves. My object has been to prepare
the ground for other labourers, by throwing on an intricate
subject all the light I could collect. I have, moreover,
attempted to trace the boundary-line between cerebral and
ganglionic affections, now considered entirely dependent
upon cerebro-spinal disturbance, largely illustrating the
varied relations in which cerebral disorders stand to ganglionic diseases." Since then, Claude Bernard discovered
that congestion, heat, and redness of surrounding tissues
resulted from the section of the cervical ganglionic nerves;
and Brown-Séquard has made us understand that this was
caused by paralysis of the vaso-motor nerves, thus giving us
the key for the right comprehension of the pathology of the
ganglionic nervous system, and enabling pathologists to
raise it from the slough of wild or silly hypothesis in which
it has slumbered up to the present time.

Since the appearance of the last edition, Dr. Beale has
published, in the Philosophical Transactions, a remarkable
paper on the microscopic anatomy of the ganglia. Dr. Handfield Jones has repeatedly made valuable contributions to our
medical journals, in which he has shown how fully one of our
best pathologists is alive to the importance of distinguishing
the diseases of the ganglionic nervous system from those of
the cerebro-spinal nerves. Dr. Edward Meryon has recently
entered the lists, and promises another valuable paper to the
R. M. C. Society's Transactions. Like these esteemed fellow-
labourers, my study of diseases of the ganglionic nervous
system, has been clinical, but I have moreover attempted to
give a general sketch of ganglionic pathology. I have not,
however, the presumption to think that I have done otherwise
than reason more or less correctly on what is known of this
recondite subject, and if those who take interest in diseases of
the nervous system will criticise this chapter to the utmost,
they will do good service to me as well as to science.

So far for the book; and now, as I have come to that
time of life when a man is frequently startled by learning

that another fellow-labourer has rested from his labour, and has for ever done with ambition, I myself wish to state in a few words what has been my object and my course of work.

Early in life I was struck by the wise remark of Goethe, that if a man wants to make his mark he must set bounds to his work, beyond the limits of which he must not suffer his steps to stray, and work to the uttermost of his power within those appointed bounds. I took to myself the physiology and the diseases of women as my field of labour— no limited speciality, I trow!

In my Elements of Health and Principles of Female Hygiene I have given the rules by which the health of women might be maintained at each successive period of life; it is a treatise of preventive medicine, for I am more and more convinced that most diseases of women are preventible, and that their frequent occurrence depends on the lamentable ignorance in which young women are brought up concerning all that relates to those very functions by which they are constituted women.

The physiology of puberty and its diseases were treated of at considerable length in the work last mentioned, and it forms the subject of the first half of my work on " Uterine and Ovarian Inflammation, and on the Physiology and Diseases of Menstruation," in which I endeavoured to show that the best way to commence the study of, and to rightly understand the diseases of women, is to study first the natural history of menstruation and its disorders. I hope at some future day to recast what I have written about this period of life, and to make a companion book to the present volume, under the title of " Puberty in Health and in Disease."

In a treatise on Uterine and Ovarian Inflammation, and in a Handbook of Uterine Therapeutics and of Diseases of Women, I record my views of diseases of women and their treatment. Among other things, I have tried to impress upon the reader that however much we may parcel out the body into more or less judiciously defined

divisions for the convenience of study and the necessities
of practice, there must be no special pathology nor therapeutics for each separate division. I have never ceased to
inculcate that medicine should be one and indivisible,
that we should reason upon diseases of the womb as we do
upon diseases of any other part of the body and treat them by
the same remedies, and that however different may be our
modes of diagnosis, they should be all based on the same
principle—that of bringing to bear every available organ of
sense upon the elucidation of disease. As far as I can judge,
this teaching has told favourably on many of the junior members of the profession. I regret, however, that I should be
sometimes consulted by women evidently suffering from
chronic inflammation of the uterus, with ulceration of the
mouth of the womb dipping into the cervical canal, as far as
can be seen, and who, nevertheless, had been told by consulting
practitioners that they were suffering from chronic debility,
and had nothing the matter with the womb. Although in
these cases the symptoms of protracted ill-health all pointed
to the uterus and to no other organ, the practitioner had
not thought it right to ascertain the nature of the case by
an ocular examination, and his diagnosis was erroneous
because he only made a digital exploration.

Although a digital examination be invaluable to ascertain
the consistency, the size, and shape of the womb, it is often
perfectly useless and unreliable for the diagnosis of ulceration, and of those various morbid states of the uterine
mucous membrane that lead to ulceration. This conviction
becomes stronger as I grow older, keeping pace with the
increase of experience, now extending over thirty years. Indeed, it cannot be too often repeated to those who treat the
diseases of women, that if the ocular examination of diseases
of the womb were to fall into disuse, all further progress of
uterine pathology would be arrested; nay more, it would
soon retrograde to what it was before Récamier took it up,
when uterine pathologists gave their chief attention to heteromorphous growths, when cancer was accepted as an ordinary

result of chronic inflammation, when all forms of acute inflammation of the womb and the ovaries were considered the same disease, and called inflammation of the bowels; while chronic uterine affections escaping detection, kept many women in a more or less constant state of ill-health; till nature fortunately came to their relief and cured them at the change of life.

I have thus stated what has been my object as well as the plan on which I have worked; and as I have already introduced this new work, I have only now to place it in the reader's hands, trusting that it may help him to solve some of the difficulties of practice; but I cannot conclude without a warm expression of thanks to my friend, Mr. JOHN CHIPPENDALE, for kindly seeing the work through the press.

E. J. T.

60, Grosvenor Street, Grosvenor Square.
September 16, 1870.

TABLE OF CONTENTS.

PART II.

GENERAL PATHOLOGY OF THE CHANGE OF LIFE.

CHAPTER III.

PRINCIPLES OF PATHOLOGY AT THE CHANGE OF LIFE.

CHAPTER IV.

PRINCIPLES OF TREATMENT AT THE CHANGE OF LIFE.

CHAPTER V.

PRINCIPLES OF HYGIENE AT THE CHANGE OF LIFE.

PART III.

SPECIAL PATHOLOGY OF THE CHANGE OF LIFE.

CHAPTER VI.

DISEASES OF THE GANGLIONIC NERVOUS SYSTEM.

CHAPTER VII.

DISEASES OF THE BRAIN.

CHAPTER VIII.

NEURALGIC AFFECTIONS.

LIST OF TABLES.

———

b

CHAPTER IX.

DISEASES OF THE REPRODUCTIVE ORGANS.

CHAPTER X.

DISEASES OF THE GASTRO-INTESTINAL ORGANS.

CHAPTER XI.

DISEASES OF THE SKIN.

CHAPTER XII.

OTHER AFFECTIONS OCCURRING AT THE CHANGE OF LIFE.

LIST OF TABLES.

b

LIST OF CASES.

LONDON MEDICAL LIBRARY

CHAPTER I.

INTRODUCTION.

THE terms "Climacteria" in Latin, "Climacteric disease," "Change of life," "Critical time," "Turn of life," in English, "Temps critique," "Age de retour," "Ménopause," in French, and "Aufhören der Weiblichen Reinigung," in German, are understood to mean a certain period of time, beginning with those irregularities which precede the last appearance of the menstrual flow, and ending with the recovery of health.

Variable as the duration of this time is in different women, it receives a certain degree of precision from the date of the last menstrual flow, which divides the change of life into two periods. When in the course of this work I speak of *cessation*, I always mean the cessation of the menstrual flow. There is no medical term to designate the time included between the first indications of the failure of ovarian energy, and cessation, but women call it "the dodging time," and as it happily expresses the uncertain and erratic appearance of the menstrual flow, I shall use the phrase to indicate the first part of the change of life in contradistinction to the last part, which begins at cessation, and concludes with the permanent restoration of health.

Although the importance of this epoch has been denied by Tissot, Dewees, Meissner, Saucerotte, and Landouzy, their opinion is not generally admitted by the profession, as is evident from many works of the last century, and from the more modern writings of Fothergill, Sir C. M. Clarke, Dr. Meigs, and Dr. G. Bedford. The last author has truly stated,

B

that " in addition to structural and malignant disorders so frequent at this period, there are many forms of eccentric nervous disturbance, various forms of temporary or permanent paralysis, and that the varieties of simple nervous irritation, without involving any peculiar lesion, are beyond calculation." German pathologists have admitted the importance of this epoch, which has, however, been chiefly recognised by the French. In addition to those whose names I shall soon have occasion to cite, Lisfranc completely adopted the popular belief in the dangers of the change of life; and Moreau de la Sarthe says, that "the change of life is characterized by headaches, syncope, leipothymia, general or partial spasmodic affections, hypochondriasis, the various symptoms of hysteria, and of insanity." Brierre de Boismont and Dr. Dusourd have recognised the vast influence of the change of life on the health of women, and I shall have frequent occasion to refer to their valuable works. I cannot equally commend the only two works expressly written on this subject, with which I am acquainted,—one by Gardanne, in 1816, entitled *Avis aux Femmes entrant dans l'Age critique*, and the other, a treatise by Dr. Menville, published in 1840, and called *Du Temps critique chez les Femmes*. These two works have little scientific merit, but their hazardous assertions have suggested inquiries which would not otherwise have been undertaken, and they have afforded me some valuable cases for the illustration of the subject.

Some admit, with Voisin, the frequency and singularity of nervous affections at the change of life, but deny that they can be caused by the monthly retention of a few ounces of blood; whereas it is contended that the cessation of the menstrual flow and the attendant changes, whether physiological or morbid, are principally induced by structural changes progressing in the ovaries, and by their various re-actions on the system. One objection claims special notice, because it is founded on an erroneous interpretation of statistics. Benoiston de Chateauneuf and Odier de Genève have proved that, if a large number of women, between the ages of forty and fifty, be compared with a similar number of men within the same periods of life, the rate of mortality will be greater among the males than the females. But those who bring forward these results, may as well say that parturition has no influence on the health of women, because it has little on their mortality; and, moreover, Finlayson has shown that, at all periods of life, more men die than women. Those who deny that the change of life is a critical period, argue as if *critical* meant *fatal*. In medical language, crisis means a sudden change for the better or the worse, leading as often to recovery as to death. Instead of flowing on in smooth tranquillity from the cradle to the grave, the stream of life is marked by rapids, which have been called critical, metamorphic, or developmental epochs, and during which an unusual predominance is acquired by one or by several of the organs which together form the human frame. The object of each successive critical readjustment of our frame is, to ensure the greatest possible amount of health consistent with each subsequent period of life. This object is attained in the vast majority of cases, but the constitution only rallies after having been severely shaken for a certain period, varying according to constitution and temperament. In some, the critical changes of dentition and puberty are brought about without ill-health, but in many, these epochs are marked by disease, or by a prolonged condition resembling convalescence, and often chequered by serious disease. At all critical epochs, the activity of some important apparatus may be too powerful, and totally disturb the functions of its allied organs, as do the reproductive organs in hysteria, or it may be too feeble to re-act with sufficient energy on the system, as is the case with the reproductive organs in chlorosis. Whether the energy of the preponderance-seeking power is above or below par, health is impaired, for the system is deprived of the power of duly reacting against stimuli, whether physical or emotional. With regard to the pathogenic influence of the critical periods of life, first and second dentition seem to influence both sexes in the same way. Puberty is common to both, but the impulse then given to the constitution of man by the sexual apparatus is, in general, fully effective and all-sufficient to ensure its permanent activity until extreme old age; whereas, in woman this crisis is very liable to be delayed or perverted, and even when puberty has been effectually established, the health of woman is dependent on those oscillations of vital power which render menstruation healthy or morbid. Matrimony, pregnancy, parturition, lactation, are like critical periods, curing some complaints, giving greater activity to others; and when, after having lasted thirty-two years, the action of the reproductive organs is withdrawn from the system, prolonged ill-health is a frequent result. Then arise a series of beautifully adjusted critical movements, the object of which is to endow woman with a greater degree of strength than she had previously enjoyed; but if the seeds of destruction have been slumbering for years in the system, the change of life will give them increased activity. Thus Dionis of old, and Madame Boivin, Dupuytren, Tanchou in our time, have proved, that the greatest proportion of cancerous affections and polypi of the womb are complained of at that period, and it is the same with cancer of the breast, as well as with adenoid, and neuromatic tumours of the breast, according to Velpeau. If the term critical be taken in a more restricted sense, as indicating a period in which the system finds relief by critical discharges, what time of life is so rife with critical phenomena? The floodings, leucorrhœa, diarrhœa, and perspirations, are eminently critical, and restore to health the vast majority of women.

This volume will forcibly show the evil effects of the change of life; its sanative influence cannot be so easily depicted, as patients consult the profession for actual disease, not to tell of diseases previously cured. With respect, however, to women who had been suffering for many years from intractable chronic affections that had baffled our best efforts to bring about recovery, the results of cessation are in general eminently satisfactory. We are most of us in the habit of telling our patients that they will be certainly cured by the change of life, and although this promise is often given to keep up the patient's hope rather than as the result of a well-grounded prognosis, still it is surprising how frequently the prophecy proves true. This remark particularly applies to ovarian congestion and subacute inflammation, and to most chronic diseases of the womb.

I have notes of some forty patients who were for many years before the ménopause confined to the bed or the sofa by chronic uterine inflammation, who made marvellous recoveries very soon after the change was effected, and who are once more actively engaged in those pleasures and duties of society from which they had been divorced for ten or fifteen years. Out of many similar cases, in which recovery was not thus rapid and perfect, I cannot call to my recollection a single instance in which great improvement was not obtained. I have also ascertained from twenty-six women who had ceased to menstruate, that they were no longer troubled by habitual leucorrhœa, and doubtless many suffer for years from unrecognised uterine affections, which are at last completely cured by the change of life. Prolapsus of the womb was cured in three cases; thirty-five women no longer suffered from uterine deviations, though they still existed. In four cases varicose veins had gone down; in twenty-four piles had disappeared, and in eight other cases they had ceased to bleed. Fifty-three women spoke of the great additional strength obtained, and of the abatement of their liability to dyspepsia. Ganglionic affections then often lose their gravity and become less frequent, and the same remark applies to almost all cerebro-spinal affections, even to the most formidable, for Esquirol has seen many women remain maniacal so long as menstruation lasted, who immediately and spontaneously recovered after the ménopause.

Reporting on the female lunatics at Colney Hatch, Dr. Davey says:—"There are many females between the ages of forty and fifty, whose recoveries may be expected when the uterus shall have fairly resumed its original inaction, and

when also the brain shall have lost a fertile source of irritation and disease.   Unfortunately it happens that the poorer
classes are much too unmindful of the health of women at
the critical periods of life, and pay too little attention to the
means whereby the uterus may be assisted in its efforts to
preserve its due influence on the human economy ; and therefore is it, in a very great measure, that insanity is of so frequent an occurrence among women."   I have dwelt on this
subject in another work,* contending that in many of the
worst cases, there was a prospect of the mental faculties
being restored at the change of life ; and quoting my suggestion, Sir W. C. Hood has admirably pleaded the cause of many
most unfortunate women confined for life at Broadmoor
for infanticide, or other great crimes, committed during a
temporary attack of puerperal mania, rightly observing that,†
though there may be danger in restoring them to society
while impregnation is possible, and therefore another attack
also of puerperal mania, there can be none in releasing them
some time after the change of life, and the late Dr. Ferrus,
Inspector-General of French Prisons, entertaining the same
opinion, the plan was successfully tried, and several women
have been restored to society, for ten or twelve years, without
a relapse having occurred.

The following case related by Négrier exemplifies the
intimate connexion between marked ovarian action and
insanity, as well as the beneficial influence of cessation :—

CASE 1.—Madame X. was tall and lymphatic, and had
been badly brought up.   Menstruation first came at twelve,
was very painful and irregular, accompanied by epigastric
oppression, palpitation, convulsions, hypochondriasis, and
suicidal mania.   During the year that followed marriage, at
thirty-eight, the attacks were less severe, but they became
more so on the menstrual flow becoming scanty, and she

---

* " Elements of Health and Principles of Female Hygiene." Bohn, York-
street, Covent-garden, London.
† "Criminal Lunacy," by W. C. Hood, M.D.   Churchill, New Bur-
lington-street.

never conceived. Some days before the menstrual period Madame X. became melancholy and secluded herself; there were lumbar and lower-limb pains, but the right iliac fossa was the spot whence radiated all her sufferings. She had made twenty different attempts at suicide, and when detected was ashamed of herself, and promised not to do it again. The fits would last five or six days, being most severe on alternate months. During the dodging time the fits were less severe, always coinciding with the menstrual flow; they failed to come in winter, and ceased altogether when menstruation left off at fifty-four.

The critical nature of a period is shown by its effects on the health in ensuing years, thus puberty is not only the crisis of most of the complaints of the preceding epochs, but it determines the health of the subsequent period for good or evil; in like manner, the change of life not only terminates critically many complaints of the preceding years, but it has a decisive action on the state of health during the whole subsequent period of life, so much so, that from the manner in which this crisis is accomplished, I believe it possible to predict whether, in after life, the health will be good or bad. Fifty-three women, in whom there had been no menstrual flow for five years, and whose health had been habitually bad, spoke of their great additional strength of constitution, and this result may be taken as a rule which harmonizes with the popular belief. It is shown by the greater longevity of women, by their being less liable to sudden death, and by their general immunity from disease, for the little said about women in works on diseases of old age is a strong proof of the good health with which they are generally blest during the last stage of existence.

In treatises on diseases of women the critical time and its infirmities are generally dismissed with half a page, and in none of our classical works have I found diseases of the change of life brought within the range of the laws of general pathology and general principles of treatment. Perhaps the deficiencies of medical literature are corrected in the daily practice, though some eminent practitioners are of a different opinion. They do not pretend that nothing is done to relieve the sufferings of women, but not enough, that a placebo is given where systematic treatment is required. Thus, Sir C. M. Clarke, commenting on the diseases of this epoch, states that " it is not unusual with women to refer all their extraordinary sensations to ' the change of life,' and to consider that, when they have thus accounted for their diseases, they have at the same time cured them ; and in this, most medical men, judging at least from their practice, seem to be of the same opinion."

Dr. Meigs likewise says : " In America, too little regard is paid to the dangers of the crisis ; and when the threatening consequences of mismanagement or misapprehension have become startling, those fatal mutations are attributed to some trivial cause, and the victim passes away to the sound of the passing bell, and no increase of knowledge, acquired by such a mournful experience, stands in the way of the next victim to a management as unwise and as thoughtless."

" The complaints," adds Dr. Meigs, " which women at the change of life often make, are frequently hushed with the unsatisfactory reply that such complaints are owing to the change of life, and are likely to cease whenever the change shall become complete. A physician has no moral right, by his opinion, to put to sleep the anxieties of his patient, and to save himself the trouble of thinking by so concise and unphilosophical a mode of proceeding. Whenever, therefore, a female, at this period, which is universally admitted to be a critical and dangerous time for her, comes to complain of symptoms referable to some morbid condition of the reproductive tissues, it is clearly our duty to give a considerate attention to her case, and not to dismiss her until our judgment should be fully satisfied as to the therapeutical or hygienical indications of the case."

So much for pathology ; and with regard to the physiological results of this great change in comparatively healthy women, I will say, with M. Floureus, that " from forty to fifty-five is a period of *invigoration* for both sexes—a period in which the deep interior work proceeding in all our tissues renders them firmer, and thereby ensures a more perfect performance of all the functions. This change is insensibly worked out in man, but in woman the passage is often perilous, and the result is more marked. The great improvement in the general health subsequent to the change of life is a notorious fact." Although the phenomena of the change of life are principally due to ovarian involution, it may be helped by other changes proceeding in both sexes. For instance, Sir W. Jenner has lately stated, in one of his remarkable lectures, " that the spleen, the lymphatic glands, and Peyer's patches all suffer involution at the same period of life—about fifty. At that time the spleen grows smaller, the lymphatic glands waste, and Peyer's patches smooth down and lose their peculiar structure."

Dr. Day, in his work on " Diseases of Old Age," has not failed to note the true character of the changes taking place at cessation, for he shows how climacteric *decay* is less frequently observed in women than in men, not only because women lead a less tumultuous life, but because their constitution has been so remodelled by the change of life that the causes of this decay have less hold over them." The immense importance of this change on the subsequent lifetime of women cannot be too highly rated, and as it is well got over or full of suffering, so will the subsequent lifetime be healthy or otherwise. It is a *final* settlement, for if it does not develop pathological seeds fatal to the system, the rest of life is generally passed in uninterrupted health, longevity being attained more frequently by women than by men. This invigoration of health is sometimes accompanied by a very great improvement of personal appearance, when bones become covered by a fair amount of fat; which "suave incrementum" is both comely and conducive to health. Others do not recover health without some sacrifice of feminine grace, their appearance becoming somewhat masculine, the bones projecting more than usual, the skin is less unctuous, and tweezers may be required to remove stray hairs from the face.

The effects of the disturbance of ovarian action on the mental faculties towards the first part of the change of life have been already alluded to. There is almost always, while the change is proceeding, a partial paralysis of ganglionic and of cerebral power, causing various forms of nervous irritability and some amount of confusion and bewilderment, which seems to deprive women of the mental endowments to which they had acquired a good title by forty years' enjoyment. They often lose confidence in themselves, are unable to manage domestic or other business, and are more likely to be imposed on either within or without the family circle. When the change is effected, the mind emerges from the clouds in which it has seemed lost. Thankful that they have escaped from real sufferings, women cease to torture themselves with imaginary woes, and as they feel the ground grow steadier underfoot, they are less dependent on others—for, like the body, the mental faculties then assume a masculine character. The change of life does not give talents, but it often imparts a firmness of purpose to bring out effectively those that are possessed, whether it be to govern a household, to preside in a drawing-room, or to thread and unravel political entanglements. When women are no longer hampered by a bodily infirmity periodically returning, they have more time at their disposal, and for obvious reasons they are less subject to be led astray by a too ardent imagination, or by wild flights of passion.

The disturbance of regular ovarian action during the first part of the change of life sometimes tells unfavourably on woman considered as a moral agent. Her mode of dealing with the every-day occurrences of life may betray a certain want of principle, contrasting in a striking manner with her previous rectitude of conduct. There is often unusual peevishness and ill-temper, sometimes assuming the importance of moral insanity. After having lived in a most exemplary manner up to the change of life, some, for a scamp, desert husband and children, while others only stay at home to make it intolerable by their ungovernable temper, and to hate the long-cherished objects of their affection. A few are most miserable pictures of melancholy self-corrosion, sitting in silent and gloomy seclusion, neither loving nor hating, only wishing to be left alone to chew the cud of most baleful introspection. Some, in the midst of affluence, indulge a propensity for stealing, and four women confessed to me that they were obliged to have their children removed, for fear they should murder them; while others are tempted to commit suicide. Unless there be some strong hereditary taint, all these shades of moral insanity are susceptible of being cured by judicious treatment, as I shall amply show hereafter, and when the change of life is passed, the habitual rectitude of moral action returns. Doubtless the subsidence of ovarian action deprives one form of love of those strong emotional impulses which gift the passion with surprising energy; but although the heart becomes at last capable of listening to the head, still love rules paramount in the breast of woman, and whether called charity, friendship, conjugal, or maternal affection, it still engrosses the thoughts, and failing all other forms and opportunities of love, religion often takes a stronger hold, crowning the evening of life with unanticipated happiness.

When safely anchored in this sure haven, a woman looks back on the time when her health was disturbed by ever-recurring infirmities, by pregnancy with its eccentricities, by the perils of child-birth, and the annoyances of nursing. From the tranquillity she has attained, she may well revert to the long years when passion, jealousy, and their attendant emotions often harrowed up her soul, presenting everything to the mind through a delusive prism. She will find how much her existence is changed from what it was, and will understand the saying of Madame de Deffand—"*Autrefois quand j'étais femme.*"

These results of the change of life are not only most important, but are daily occurring in about 1,200,000 of our countrywomen, so I need offer no apology for studying their complaints more accurately than has been previously done; and as I am one of those who maintain that physiology and pathology are interdependent, that their rate of progress invariably coincides, I infer that the little knowledge of diseases of the change of life is principally owing to the neglect in which physiologists have left many questions relating to this epoch, and that a better understanding of it is not to be attained without a careful investigation of all that relates to its physiology, and this I shall attempt in the next chapter.

PART I.

PHYSIOLOGY

OF

THE CHANGE OF LIFE.

CHAPTER II. .

PHYSIOLOGY OF THE CHANGE OF LIFE.

If, hitherto, some of the physiological phenomena of the change of life have been unnoticed or imperfectly understood, it is because physiologists have not sufficiently considered their relation to the other critical periods of the reproductive function. If pathologists are at a loss how to treat the continually recurring infirmities of the change of life, it is because they have not compared them with the diseases frequently met with at puberty. To explain how the change of life so completely modifies the constitution of woman, leading either to improved health or to diseased action, I am obliged rapidly to review the reproductive function, referring those who wish for further details to another of my works.*

About the age of thirteen, in the latitude of Great Britain, the constitution of girls is seen to change. Those who are delicate become more so ; those who are in rude health suffer slightly. There is a frequent recurrence of pains in the back, in the lower limbs, and in the abdomen ; there is a tendency to headache, nervous irritability, and fitfulness of temper. After this has lasted for twelve or eighteen months, a small quantity of blood is perspired from the womb, and the nervous symptoms abate, but they generally return again for a few months, until the recurrence every twenty-eight days of the menstrual flow, with a certain amount of nervous irritability, pain, and headache, has become a confirmed habit. In the

---

* "On Uterine and Ovarian Inflammation, and on the Physiology and Diseases of Menstruation." 3rd edition.

mean time the body attains to its full proportions; the pelvis
acquires sufficient size to permit the passage of a child; the
breasts swell to prepare its food; there is a general improvement in health, and woman soon attains to her highest point
of physical perfection.   Unless interrupted by pregnancy and
lactation the menstrual flow recurs every month for about
thirty-two years, then it becomes irregular as at puberty, and
after being so for about two years it ceases altogether.   As
at puberty, so at the change of life, women are habitually
more irritable, more sensitive, more subject to headaches
and to lowness of spirits.   As at puberty so at the change
of life, a renewal of strength generally follows.   Such is the
menstrual function under favourable circumstances, but there
is a reverse to every medal.   Instead of recognising puberty
by increased health, how often do we infer it from the un-
developed form, the sickly look, and the chlorotic cast of
countenance.   Instead of slight pains in the back, limbs, and
hypogastric regions, the pains may be severe and long continued, leading to, what has been called, spinal irritation and
contraction of the limbs.   Headache may incapacitate the
patient for all exertion; her mental faculties may seem lost
in a state of drowsy forgetfulness or pseudo-narcotism; nervous irritability may grow to hysteria under its worst forms;
and instead of fitfulness of temper and waywardness of disposition, she may become morose and mischievous.   In fact,
the conduct of a girl at puberty sometimes betrays such a
dereliction of all principle, that moral insanity is not too
strong a term to characterize it.   When the long-expected
menstrual flow at last appears, it does not always bring
health.   Its return may be too long delayed, or should it
come regularly, instead of being a bearable infirmity, it may
be attended by so much abdominal pain and nervous symptoms as to constitute a disease; or there may be something
amiss in the reaction of the reproductive organs on the nervous system, so that the menstrual flow may return every
fifteen or twenty-one instead of twenty-eight days.

The change of life has also its diseases.   Instead of being a period marked by slight infirmities, it may be characterized by very complicated morbid phenomena. There may be great debility and chlorosis as at puberty, or confirmed biliousness; the abdominal pains may be very severe; headaches frequent and agonizing; instead of a slight haziness of the mental faculties there may be continued drowsiness and stupor; if hysterical convulsions are less frequent than at puberty, the globus hystericus and the minor manifestations of hysteria are very frequent, and insanity occurs oftener. If unrelieved by repeated critical discharges, such as flooding, leucorrhœa, diarrhœa, or perspiration, these complicated nervous symptoms may continue for years, and still be amenable to proper treatment, by which their occurrence might have been effectually prevented.

Puberty and the change of life are caused by anatomical changes, the one by ovarian *evolution*, the other by ovarian *involution*. At puberty the ovaries increase in size, become more vascular, and let fall ovula. At the change of life ovula are more and more scantily secreted. Little is known of the anatomical condition of the ovaries *during* the dodging time, and those in hospital practice would do well to ascertain whether they present any particular appearance in women dying at that epoch. After cessation, from smooth and turgid the ovary becomes shrivelled up, so as to resemble a peach-stone; then it becomes more and more difficult to trace the cavities of the Graafian vesicles, for their hypertrophied walls are pressed together. Later in life it is more and more atrophied, and I have found it not larger than a horse-bean, the place of the ovary being sometimes only indicated by a small fibro-cellular substance. This ovarian atrophy determines corresponding changes in the Fallopian tubes, which contract and are sometimes obliterated; it also causes the womb to become atrophied, and its neck thinner and shorter. Virchow has noted the closing of the internal os uteri in old age, and I have found an obliteration of the os uteri in five women, whom I had occasion to examine during life, from ten to fifteen years after cessation. It is

c

well known that the vagina often becomes narrower and shorter, and the pampiniform plexus—a vascular contrivance to supply the organs of generation with a large amount of blood—shrivels up. The mammary glands are likewise atrophied after cessation, whereas during the change of life they are often congested and painful, if not otherwise diseased.

It will ever be a matter of surprise how so many phenomena of health and symptoms of disease can be determined by two little oval bodies, whose structure does not appear to be complicated, but as the fact is unquestionable, it is well worth attempting to explain the varied manner in which these organs influence the system in health and disease. The ovaries influence the various parts of the body through the medium of their nerves, for as they have both ganglionic and cerebro-spinal nerves, they can re-act on both the ganglionic nerves and their centre, and on the cerebro-spinal nerves and their central organs. Whether the ganglionic be considered an independent system of nerves or an offshoot of the cerebro-spinal nervous system, all agree that it follows every capillary to its minutest ramifications, is vaso-motor, and governs the nutrition of all and every part of our frame. It is known that all important organs of nutritive life are supplied with ganglia and a plexus of ganglionic nerves, that they all communicate together, and with a larger plexus and more voluminous ganglia, situated at the pit of the stomach, called the solar ganglia and plexus; and before these foci of nervous matter had been noticed, Galen had called the epigastric region the lever of the forces by which the animal economy is moved. Without entering into details, which will be found in a subsequent chapter on ganglionic diseases at the change of life, I shall merely state, that while sensation and motion are intimately dependent on the cerebro-spinal system, nutrition is dependent on the ganglionic, and that there is a condensation of ganglionic nervous power in the central ganglia, which gives and receives from each viscus a variable impetus. Many pathological facts, which I shall set in order, are not to be accounted for without taking it for granted that the solar ganglia form an important centre of nerve force, capable of controlling and of disturbing the various parts of the body, by some subtle invisible fluid analogous to electricity, which I shall speak of as nervous fluid. Reasoning, then, from the facts to which I have alluded, it seems that the human frame is so constructed, that its various component organs can re-act upon each other in the way most conducive to health, until the age of puberty. At that time health often fails, and the whole system languishes, unless the reproductive organs come into full activity. From puberty until the change of life, the health of woman cannot be maintained without an appropriate amount of ovarian influence. If this ovarian aura re-acts healthily, it augments the power of the epigastric nervous centre, causes the functions of nutrition to be performed with increased energy, and gives an instinctive consciousness of strength. If, on the contrary, this ovarian aura be insufficient in quantity or defective in quality, it half-paralyses the epigastric centre of ganglionic action, and uneasy sensations will be felt at the pit of the stomach,—feelings of sinking and faintness, actual fainting being sometimes induced. There is often pain, or anomalous sensations; and frequently a state of defective nutrition or chlorosis, a disease doubtless generally met with at puberty, but also of occasional occurrence during pregnancy and lactation, and which is often well marked at the change of life. If the ovarian stimulus be insufficient, it will retard the first appearance of menstruation, cause the flow to come irregularly, or for a time suspend its recurrence. If the ovarian stimulus be too strong or defective, it re-acts with morbid energy on the other abdominal viscera, likewise endowed, each with its special ganglionic plexus and ganglia. It is not surprising that organs similarly endowed should sympathize, and that whether at each menstrual period, at puberty, or at the change of life, undue ovarian influence should give rise to more or less nausea and sickness, or to a perverted appetite; that it should cause the intestinal

canal to secrete more gas or mucus than usual, thus determining diarrhœa ; that it should force the lower part of the
bowel to perspire blood ; increasing the amount of water
passed by the kidneys, and of the sediments contained in the
urine, and rendering more abundant the amount of saline
water perspired by the skin. If at each menstrual epoch, or
at the beginning or cessation of the reproductive functions,
the ovarian stimulus be too abundant or too energetic for
the allied abdominal organs, then may also arise painful or
strange sensations in the ovarian regions ; for at each menstrual crisis, during puberty or the change of life, the spinal
and hypogastric pains are caused by undue ovarian influence
acting through the splanchnic and the spinal nerves. Some
amount of pain is perfectly consistent with the healthy performance of the menstrual crisis, with its inauguration at
puberty, and with its demise at the change of life, for when
the ovarian influence is defective, as in chlorosis, there is
often no pain. The too abundant or too powerful ovarian
action, after determining pain in the ovarian regions, causes
pain, oppression, or uneasy sensations in the epigastric region,
and sometimes globus hystericus ; but its area of action is
not confined to ganglionic nerves, or to nerves intimately
blended with them, for through the medium of spinal nerves
the ovarian stimulus acts on the spinal cord, and so develops
its energy, that, whether at the monthly ovarian crisis, at
puberty, or at the change of life, there arises frequently a
tetanoid condition of the system, a state of nervous irritability,
shown by an impossibility of repose, by a continued state of
restlessness, by the fidgets, by what a French writer has
called "impatiences organiques," or by hysterical convulsions.
Or there may be numbness of some part of the skin, local
anæsthesia, local paralysis, called hysterical in young women,
and occasionally paraplegia or hemiplegia. The strength of the
ovarian nisus, or the relative weakness of the nervous system,
may be inferred from the fact of death occurring in the midst
of hysterical convulsions, without any lesion being detected in
the cerebro-spinal system. The ovarian influence having

reached the ganglionic centre, is in communication with the brain by the pneumogastric nerves, so that at each menstrual epoch, at puberty, or at the change of life, its undue influence may be shown in some by distressing headaches, continued fretfulness, peevishness, and capriciousness, called temper, by a temporary perversion of moral feeling, or by moral insanity. In others, excessive ovarian action is manifested by motiveless high spirits or depression, by delirium, then called hysterical, and very frequently there is more or less of what I have termed pseudo-narcotism, in which case the cloud is no longer on the moral instincts but weighs on the mental faculties, which are, for a time, in a state of misty haziness, the brain feels muddled, memory is faithless, there is an unconquerable desire to sleep during the day, even when the night has been passed in repose. This state of the nervous system has in some cases amounted to coma and lethargy.

Having thus briefly sketched the varied phenomena which attend on the successive stages of the reproductive function, I am able to open the investigation of its terminal crisis.

To ascertain the date of cessation is a point of great importance, for it often marks the climax of much previous and subsequent suffering. In the following table I have placed side by side, the results obtained in Paris by B. de Boismont, and in London by Dr. Guy and myself. It will be seen that although the date of cessation varies from the twenty-first to the sixty-first year, yet it may generally be expected from forty to fifty. Thus, out of B. de Boismont's 181 cases, in 114, cessation took place between forty and fifty inclusively; in 330 of my 500 cases, cessation occurred during the same decennial period : and out of Professor Hannover's 312 cases of cessation in Denmark, 250 occurred during the same period of time.

TABLE I.

Dates of the CESSATION of Menstruation.

Age at Cessation.	Paris. B. de Boismont's cases.	London. Dr. Guy's cases.	London. Dr. Tilt's cases.	Total cases.
21st year	2	...	...	2
24 ,,	1	...	...	1
25 ,,				
26 ,,	1	...	...	1
27 ,,	1	1	1	3
28 ,,	1	1	...	2
29 ,,	1	...	...	1
30 ,,	...	1	10	11
31 ,,	3	...	1	4
32 ,,	2	...	4	6
33 ,,	...	2	1	3
34 ,,	4	1	2	7
35 ,,	6	3	6	15
36 ,,	7	1	2	10
37 ,,	4	5	7	16
38 ,,	7	5	6	18
39 ,,	1	7	10	18
40 ,,	18	33	42	93
41 ,,	10	24	17	51
42 ,,	7	24	26	57
43 ,,	4	23	24	51
44 ,,	13	24	23	60
45 ,,	13	45	49	107
46 ,,	9	34	31	74
47 ,,	13	25	42	80
48 ,,	8	38	37	83
49 ,,	7	25	32	64
50 ,,	12	37	49	98
51 ,,	4	14	27	45
52 ,,	8	13	16	37
53 ,,	2	8	9	19
54 ,,	5	2	7	14
55 ,,	2	1	6	9
56 ,,	2	2	4	8
57 ,,	2	1	2	5
58 ,,	...	...	4	4
59 ,,	...	...	1	1
60 ,,	1	...	1	2
61 ,,	...	...	2	2
Total	181	400	501	1082

Average date of Menstruation.

Years	44·0	45·8	46·1	45·7

It will be noticed that the average date of cessation is higher in proportion to the amount of cases, that the average of the 1082 cases is forty-five years nine months, and that more women cease to menstruate in the fiftieth year than in any other, which is confirmed by Mayer's experience at Berlin, and by that of Lieven, at St. Petersburgh.

What is known of the cessation date of menstruation is shown by the following table :—

TABLE II.

Comparative Dates of Cessation in different Countries.

| Countries. | France. | | England. | | Central Germany. | Denmark. | Norway. | Lapland. | Russia. |
	Paris.	Rouen.	London.	Manchester.					
Number of Cases.	178	190	500	1586	824	312	391	34	100
Average Date of Cessation.	44·0	48·7	46·1	47·5	47·0	44·8	48·9	49·4	45·9
Observers.	Brierre de Buismont.	Leudet.	Tilt.	Whitehead.	Mayer.	Hannover.	Faye and Vogt.	Vogt.	Lieven.
Sources of Information.	Traité de la Menstruation.	Paris Medical Congress.	This Work.	On Abortion and Sterility.	Paris Medical Congress.	Work on Menstruation.	Paris Medical Congress.	Paris Medical Congress.	Paris Medical Congress.

It is difficult to understand why there should be a difference of nearly five years between the dates of cessation of 312 Danish and 391 Norwegian women, but the similarity of results obtained by Professor Faye and Dr. Vogt should suggest to my friend Dr. Hannover to test the date of the ménopause in Denmark, by more extended figures.

There is nothing positively known respecting the date of cessation in tropical climates. Are we to believe the Hindoo medical student, who, out of his own circle of acquaintance, gave Dr. Webb* a list of thirteen Hindoo women in whom cessation occurred at the following very late dates ?

* "Pathologica Indica," Part II. p. 279.

Years.		Cases.	Years.		Cases.
50	. . . . . .	1	63	. . . . .	1
56	. . . . . .	2	64	. . . .	1
57	. . . . .	1	65	. .	1
58	. . . . . .	1	67		1
59	. . . . . .	1	68	. .	1
60	. . . . . .	1	80	. . . .	1

I could not match this list, from amidst those who consult me, nevertheless, a comparatively late date of cessation in India is rendered probable by the fact that Hindoo legislators place the ménopause at fifty, and in the Koran, women are only considered too old to have children after the fifty-third year of their age.

Having thus established the average age of cessation, the frequent exceptions to the rule remain to be considered, bearing in mind that it may be taken for granted that late fecundity proves late menstruation. Twice have I known the menstrual flow to continue its regular appearance up to the sixty-first year, in ladies of a remarkably strong constitution : which was the case with a lady who regularly menstruated up to the time of her death in her eighty-fourth year.

Mr. Roberton observes :—" I am able to speak confidently concerning three women who had children at advanced ages,—one in her fiftieth year, another in her fifty-first, and the third in her fifty-third year. In each of these instances the menstrua continued up to the period of conception." He also cites a case where menstruation ceased for twelve months about the fiftieth year, when it again became regular and continued so until the seventieth. Mr. Davies, my colleague at the Farringdon Dispensary, has confined a lady of her thirteenth child, at the age of fifty-three, after which there was no menstrual flow ; and Dr. Davies published the case of a woman who was fifty-five when her last child was born, and who menstruated up to conception ; and Dr. Meyer, of Berlin, in his remarkable paper* mentions his having ascer-

---

* Congrés Médical International de Paris.

tained that out of 6000 women, menstruation was still progressing in twenty-eight at the age of fifty, in eighteen at fifty-one, in eighteen at fifty-two, in eleven at fifty-three, in thirteen at fifty-four, in five at fifty-five, in four at fifty-six, in three at fifty-seven, in three at fifty-eight, in one at fifty-nine, in four at sixty, in four at sixty-two, and in three at the age of sixty-four, the seven last cases occurring in the upper classes. These unusually protracted dates make it appear singular that Hannover should have observed nothing similar in Denmark, for he only notes the exceptional occurrence of cessation in one case at fifty, two at fifty-one, three at fifty-two, one at fifty-three, and one at fifty-four.

Dr. Marion Sims has met with two cases of parturition at fifty-two, and a negro woman was said to be sixty, when she was confined twenty years after her last confinement.

Lamotte relates that a woman had thirty-two children, and menstruated quite regularly up to her sixty-second year. Auber attended two women, one sixty-eight and the other eighty, who for the last few years had again menstruated. The flow came regularly, lasted three or four days, and during that time they were more nervous than usual, the organs of sensation being unusually dull in apprehending their appropriate stimuli. Saxonia states that a nun, in whom the menstrual flow ceased at the usual time, experienced its return when her 100th year was attained, and it continued regular until her death, three years after. Rush mentions the case of a woman who was confined for the last time in her sixtieth year, menstruated until her eightieth, and died in her 100th year.

Haller records two cases in which women at sixty-three and seventy respectively bore children. Capuron cites the case of a lady, who after the menstrual flow had been absent for several years, saw it return at sixty-five. Three months after she miscarried, the fœtus being well formed. Meissner states that a woman who first menstruated at twenty, bore her first child at forty-seven, and the last of seven other children at sixty. Menstruation ceased and reappeared at seventy-five, continuing until ninety-eight, then stopped for five years, again to return at the advanced age of 104. In 1812 she was still alive.

I might increase the number of such cases, which are not instances of irregular flooding, but of the menstrual flow, occurring regularly with its attendant symptoms, or followed by pregnancy. These facts contradict the opinion of those who assert that when the menstrual flow has once fairly ceased between forty and fifty, any blood that may afterwards flow from the womb must depend upon some undetected ulceration, but in most of the cases of late menstruation that have come under my notice, I have not found any ulceration.

The following table, extracted from the Registrar General's Report, shows that many more women than is supposed bear children late in life:—

TABLE III.

Fecundity of Women at various Ages.

Ages of Mothers when their Children were born . . .	Under 20	20—25	25—30	30—35	35—40	40—45	45—50	Above 50	Total.
Children born from 1831 to 1835 . . .	8301	70,924	121,781	126,808	98,950	49,660	7022	167	483,613

Thus, out of 483,613 women who became mothers from under twenty to above fifty, no less than 7022 bore children from their forty-fifth to their fiftieth year, and 167 were mothers after having passed their fiftieth year; that is to say, that 7189 women out of 483,613, or one in about sixty-seven, bore children after the time when the menstrual flow usually ceases.

The Irish fecundity table to which Dr. Routh refers in a paper on procreative power,* is equally conclusive :—

---

* "London Journal of Medicine," 1850.

TABLE IV.

*Table of the Number of Marriages, and the Issue from such Marriages,
occurring in the Eleven Years ending* 1841, *for Women of the Ages
under* 17, 46-55, *and above* 55.

Age of Husband.	Age of Wife.	No. of Marriages.	Prop. of Children to 100 marriages.	Age of Wife.	No. of Marriages.	Prop. of Children to 100 marriages.	Age of Wife.	No. of Marriages.	Prop. to 100 Marriages.
Under 17		661	256		1	0		1	0
17—25		9847	262		35	51		3	0
26—35	Under 17.	4066	252	46 to 55.	145	51	Above 55	12	0
36—45		313	248		227	39		15	20
46—55		36	205		428	22		52	10
Above 55		18	128		295	10		136	12
Total ...		14,947	257		1131	26		219	12

Thus it appears that in ten years, out of 427,977 women
married in Ireland at different ages, there were no less
than 1131 married between their forty-sixth and fifty-
fifth year, and 219 were married after their fifty-fifth year.
Those who wish for further evidence of the fact under
discussion may consult the Registrar-General's Seventeenth
Report.

The following table shows similar results in a northern
latitude :—

TABLE V.

Fecundity Table for Sweden and Finland.

Age.			Annual Average Number of Deliveries.	Proportion to 100 Females living.
15	to	20	3,298	2·48
20	—	25	16,507	12·56
25	—	30	26,329	21·64
30	—	35	25,618	22·82
35	—	40	18,093	18·32
40	—	45	8,518	9·54
45	—	50	1,694	2·28
50	—	55	39	·05

In the paper referred to, Dr. Routh has published corresponding results derived from two London institutions. At
the St. Pancras General Dispensary, out of 1527 women

confined, the age of twelve varied from forty-six to fifty-seven. Two of the twelve were fifty-five and one fifty-seven. At the Westminster General Dispensary, out of 2509 women confined, the age of six varied from fifty-six to forty-nine. Dr. Taylor* has given the following table, drawn up by Neverman, which shows how suddenly the number of pregnancies diminishes after the forty-fifth year :—

TABLE VI.

Of 10,000 pregnant women, 436 or $43\frac{1}{2}$ per 1000 were upwards of 40 years of age. Of these 436 women,

101 or $10\frac{1}{10}$ per 1000 were in their 41st year.

113 or $11\frac{7}{10}$	,,	,,	42nd ,,
70 or 7	,,	,,	43rd ,,
58 or $5\frac{4}{5}$	,,	,,	44th ,,
43 or $4\frac{3}{10}$	,,	,,	45th ,,
12 or $1\frac{1}{5}$	,,	,,	46th ,,
13 or $1\frac{3}{10}$	,,	,,	47th ,,
8 or $\frac{4}{5}$	,,	,,	48th ,,
6 or $\frac{3}{5}$	,,	,,	49th ,,
9 or $\frac{9}{10}$	,,	,,	50th ,,
1 or $\frac{1}{10}$	,,	,,	52nd ,,
1 or $\frac{1}{10}$	,,	,,	53rd ,,
1 or $\frac{1}{10}$	,,	,,	54th ,,

Judging from many of the cases just recorded, fecundity was as remarkable as the protraction of the menstrual flow; and if so many women bore children after forty-five, it was not because they had married late, but because their ovaries had been endowed with unusual vitality. This view is supported by the experience of Mr. Roberton, who states "that in eleven women, three had a child each in the forty-ninth year, and the other eight had each a child above that age;

* "Medical Jurisprudence," p. 568.

I ascertained that the aggregate number of their children was 114—*i.e.*, ten and a fraction for each woman, a fact indicating that they must have married rather early in life. Concerning the age of marriage in two out of the eleven, I possess some little information; the one married at eighteen, had two children before she was twenty-one, and brought forth her fourteenth child in her fiftieth year; the other was married from a boarding-school at a very early age. In her fifty-third year she was delivered of her twelfth child."

Protracted menstruation also is proved by the following table :—

TABLE VII.

Ages of Women dying from Childbirth in the Year 1852 in England and Wales.

Ages.				Deaths from		
				Metria.		Childbirth.
15 and under	20			50	...	80
20	,,	25	...	210	...	340
25	,,	30	... ...	231	...	461
30	,,	35	...	217	...	486
35	,,	40		173	...	544
40	,,	45	...	78	...	313
45	,,	50		13		51
All Ages				972	...	2275

I have insisted on the frequency of pregnancy late in life, because grievous mistakes have often followed the practitioner's persuasion of its impossibility. To my knowledge, pregnancy late in life has been mistaken in three cases for an ovarian tumour, and was treated by iodine, mercurials, and tight-bandaging, which caused the death of the child, and greatly compromised the mother's health. The popular belief, that parturition is more dangerous to the mother when it occurs late in life is confirmed by Dr. Matthews Duncan, who has made out that if a woman has a large

family, she only escapes the extraordinary risk to be encountered at a first confinement to encounter another extraordinary risk at a ninth or subsequent labour, and his
statement is confirmed by Dr. Hugenberger of St. Petersburgh.

It will be gathered from the preceding tables that
the average date of the ménopause does not correspond
with the last parturient efforts of woman, and Mr. Whitehead has brought this out by operating on 1586 cases,
and he found that while the average date of last menstruation was 47·5; the average date of the last confinement
was 41·7.

What are the causes of protracted menstruation? The
ovaries may become paralysed before the time usually fixed
for their atrophy, and resume their wonted energy by a
spontaneous effort, by the shock of sudden grief, or through
some impulse given to the ganglionic system, by fevers and
visceral diseases. At the same time it is obvious, that a
sanguineous discharge from the womb must not be accounted
menstrual, unless it be repeated regularly; for fever may
cause menorrhagia, as it may epistaxis. For instance, Gardanne gives a case, wherein an abundant menstrual flow is
said to have come, for the last time, after six months' stoppage, to a woman forty-nine years of age, during a bilious
fever, in which emeto-cathartics were given. This might
have been uterine hæmorrhage: so might it have been in
Bohnius' case, where the ingress of fever is stated to have
brought on a return of menstruation in a woman eighty years
old. The same remark applies to a statement lately made by
Mr. Wood, of the return of the menstrual flow in a lady
aged sixty-nine, in consequence of the death of a favourite
son; and at sixty, to her sister, in consequence of a fright.
In like manner I have three patients aged fifty-two and
upwards, very much subject to heats and flushes, and in
whom worry and excitement cause the womb to bleed. Such
cases are open to doubt, unless the critical discharge returns
regularly for a certain time, as in the following instance

published by Mr. A. Brown.* A woman had not menstruated since her forty-second year, when, after suffering seven months from swelling of the liver and pains in the loins, she was critically relieved in her fifty-sixth year, by the sudden appearance of menstruation, which was repeated ten times, and perfectly re-established her health. Protracted menstruation is, however, more frequently caused by affections of the womb than of any other organ, and fibrous tumours of the womb often retard the date of cessation. Uterine polypi have the same effect, and in some of Dupuytren's cases of uterine polypi the menstrual flow lasted until the forty-ninth or fifty-sixth year; even in these cases the sanguineous discharge should not be considered menstrual unless it occurs periodically, or with periodical paroxysms. I have sometimes found that hypertrophy and ulceration of the neck of the womb or other uterine diseases coincided with an unusually protracted menstrual flow, and as the earlier observers had not the means of recognising them, their cases of late cessation are, to a great extent, invalidated. I repeat that there may be uterine bleeding without menstruation, and I do not consider as cases of menstruation the case of a woman, who at seventy had an attack of apoplexy, and a red discharge from the vagina, the uterus was found full of blood, and the tubes and ovaries much injected. In another case likewise reported by Andral, a woman died at seventy-five of hemiplegia from extravasation in the left side of the cerebellum, the womb was full of blood, and its tissues intensely red.

With Dr. Dusourd, I consider ovarian activity to be commensurate with constitutional vigour, inasmuch as all those in whom I have observed the menstrual flow to be unusually prolonged, were remarkable for their strength and good health. Neither should the philosopher lose sight of the connexion between the unusual prolongation of ovarian life and longevity, remarkable in several of the preceding cases, and I believe that life is longest in those women in

---

* "London Medical Gazette," vol. xxi.

whom puberty is retarded, as it is proved to be the longest
in cold countries where the average date of first menstruation is delayed beyond the average in temperate climates.
Alexander von Humboldt has arrived at the same opinion,
founding it upon extensive and comparative study of the
numerous races which inhabit South America, and it
may be assumed that when the ovarian nisus is healthily
manifested, it indicates a corresponding healthy activity
of the other functions of vegetative life; and when it is
unusually prolonged, it implies a corresponding power of
endurance of vegetative life, on which depends longevity,
and the frequent protraction of cessation to the forty-
eighth and the fifty-fourth year, which is said by Dr.
Robert Cowie to occur in the Shetland Islands, corre-
sponds with the frequent occurrence of longevity in the same
islands.

Another means of judging whether in a temperate climate
cessation will be retarded is the circumstance of its having
first appeared later than usual. In some of the cases of very
prolonged menstrual flow, it first appeared as late as twenty
and twenty-two. On comparing thirty-three women, who had
first menstruated from eight to eleven, with thirty-seven
women, in whom menstruation had been retarded from the
eighteenth to the twenty-second year, I obtained the results
shown in the following table :—

TABLE VIII.

Date of last Menstruation in the EARLY *Menstruated and the*
LATE *Menstruated.*

Years of Age at Cessation.	Cases of early Menstruation at from 8 to 11.	Cases of late Menstruation at from 18 to 22.
30	2	1
31	1	
34	1	
35	1	
37		1
38	1	
39	2	1
40	2	3
41	1	1
42	2	1
44	2	2
45	3	3
46	1	7
47	2	2
48	1	3
49	3	3
50	4	4
51	1	2
52	1	
55	1	1
58	1	2
Total	33	37

| *Average date of Cessation.* | | |
| Years | 44·6 | 46·8 |

Two years and three months is a remarkable difference
between the two averages, 45·7 being the average of the date
of cessation in 1082 women.

These numbers are too small to be conclusive, but they tell
against the contrary opinion of Raciborski, who seems to
have been too strongly impressed by very exceptional cases,
like that cited by Descuret, of a woman who first menstruated
at two years of age, married at twenty-seven, had a numerous
family, and menstruated regularly up to her fifty-third year.

Supposing, however, the table to be true for England, it is
not so in Denmark, where Hannover found menstruation to
cease soonest in those women in whom it began latest; and
Dr. Mayer's figures confirm the assertion of Burdach and
Mende, that the same holds good in central Germany.
Thus, while out of 722 women, 13 per cent. who first menstruated from eleven to thirteen ceased to menstruate at
fifty; 11 per cent. who first menstruated from eighteen to
thirty-one ceased at forty-seven.

With regard to the structural conditions, whereby in a
certain number of women, the menstrual flow and fecundity
are protracted to such an advanced age, it is fair to suppose
that they were anatomically constituted like a lady in whom
menstruation only became regular at the last years of her
life, and who died at seventy-two, and in whom Drs. Bouvier
and B. de Boismont found the ovaries and the whole of the
generative system turgid, as in girls of fifteen to eighteen
years of age; instead of being in their usual state of atrophy.

I now come to the causes of early cessation, or premature
ovarian paralysis. As human life is often cut short long
before the time specified by the Psalmist, so the reproductive
force with which the ovaries are endowed may be extinguished
long before the average date of forty-five. On careful consideration of forty-nine cases, wherein menstruation ceased
suddenly from the twenty-seventh and thirty-ninth year inclusively, I was unable to detect anything peculiar to their
constitution with the exception of eight, whose strength was
below the average. None of the forty-nine had married too
early or were addicted to prostitution, which have been
erroneously given as causes of early cessation by Meissner.
They were not unusually subject to profuse menstruation, as
asserted by Gardanne. They were not more than usually
affected with disease of the reproductive organs, as stated by
Meissner, neither did the diminished extent of the menstrual
function indicate less reproductive power, for the average
amount of children was a little more than three in the twenty-
six out of the forty-nine women who were married.

If women of the same family sometimes cease to menstruate at the same age, it may be a coincidence, but I have repeatedly known sisters to cease to menstruate at the same time of life. I attend two out of seven sisters; they all first menstruated from the tenth to the thirteenth year, and in all menstruation ceased about the fortieth year. Dr. Mayer has found, on comparing women of the lower and higher classes, that menstruation came later and ceased earlier in the lower classes, while in the upper classes menstruation came sooner and ceased later.

Out of the forty-nine cases in which the menstrual flow was suppressed from the twenty-seventh to the thirty-ninth year, in twenty-seven I found sufficient cause to explain the fact, as will be seen from the following table :—

TABLE IX.

Causes of early Cessation in 27 Women.

Nature of Cause.	No. of Women
Parturition and lactation	3
Miscarriage	1
A fall on the sacrum during menstrual flow .	2
Suppression of menstrual flow, from frost or intense cold	2
Bleeding from the arm, at a menstrual period .	1
Getting wet through, during a menstrual period	1
Violent purging, the result of medicine . .	2
Cholera	2
Rheumatic fever	2
Bronchitis, with fever	2
Fright	9
Total	27

Some of these cases deserve a few words of comment. Lucy A., aged fifty-three, had never menstruated since a miscarriage at thirty-two, and her health has been better ever since.

W. N., now fifty, a tall, stout-looking woman, had a child at thirty-four, continued to flood for eight weeks. She suckled her child for a year, but has never menstruated since.

Elizabeth A., a woman of average strength, aged forty-seven, had a child ten years ago, suckled it eighteen months, and menstruation did not return. In such cases ovarian energy seems exhausted by parturition and lactation, and nature seizes the opportunity to put a sudden end to the menstrual function.

Julia L., a chlorotic-looking woman, aged forty-five, had no menstrual flow since she was thirty-three, when during menstruation, she missed a few steps and fell on her back. Here the ovaries were so paralysed by a mechanical shock that they never recovered.

Emily F., a healthy-looking person, aged forty-eight, last menstruated at thirty-nine, when the menstrual flow was suddenly stopped by severe frost. She often had leucorrhœa during the two years which followed, but did not otherwise suffer. Another woman, aged thirty-four, went out shopping on the first day of menstruation, the weather being intensely cold; the flow was suddenly stopped and never returned, health remaining uninterruptedly good. In such cases cold has the same paralysing influence as a fall, and similar effects may be produced by cholera or violent purging. Dusourd relates three cases where the menstrual flow ceased after the supervention of diarrhœa, coming about every month and lasting for a few days; and he has known the menstrual flow to cease in three women from forty to forty-three, after an abundant hæmorrhoidal flow. Fever has so powerful an effect on the viscera principally endowed with nervous energy by the ganglionic system, that one can understand why it should be a cause of ovarian paralysis.

As at puberty the menstrual flow is sometimes suddenly brought on by emotion, so fright and the sudden communication of bad news are the most frequent causes of sudden early cessation. The ovaries and womb are stunned by a nervous shock like the frog's muscles by a strong electric shock.

Ann G., aged fifty-nine; the menstrual flow ceased at thirty-nine, being suddenly suppressed by grief for the death of her husband. The same occurred at thirty-four to Maria B., now aged fifty. Jane O., a healthy woman, had a flooding at forty-five, while witnessing an execution, and the menstrual flow never returned.

Elizabeth C., a dark-eyed, thin, sharp, and wiry old woman of seventy-one, had enjoyed excellent health until her thirtieth year, and had been nursing a child sixteen months, when her husband suddenly dropped down dead at her feet. Stunned by the blow, she let the child slip from her arms, lost her senses for a few hours, and her milk "turned to water." For a time she felt a good deal shaken, but was always able to attend to her household duties; and her health soon became as strong as usual and continued so until lately, although the menstrual flow had not appeared since her twenty-eighth year, and although the system had not been relieved by leucorrhœa, diarrhœa, piles, or by systematic treatment. A woman had been quite regular up to the age of thirty, when she threw a dead rat on the fire, thinking it was a coal; she retched for hours; for four years there was no menstrual flow, and after that an occasional show, at irregular times.

Eliza S., aged fifty-two, had been regular from eighteen to thirty-nine, when she was bled at a menstrual period for rheumatic pains in the shoulders. Immediately after being bled she vomited for an hour, had intense perspirations, but never menstruated again.

Ovarian tumours often cause early cessation of the menstrual flow. Thus, out of fifteen cases of ovarian tumours operated on by Dr. W. Atlee, of Philadelphia, some referred to young women, but in most cases the amount of the menstrual flow was diminished, and cessation occurred in four at thirty, thirty-nine, forty, and forty-two. I have seen internal metritis cause cessation at twenty-nine; but I do not agree with Scanzoni, that premature cessation is to be expected as the ordinary result of internal metritis, neither does my experience agree with that of Dr. Emmet, who thinks that early cessation is caused by Dysmenorrhœa, partly obstructive and partly inflammatory; but I have repeatedly known cessation to occur about thirty, when it had been habitually scanty, as if the system found it too great an effort to continue the flow.

Diagnosis of Cessation.—Every well marked condition of the constitution has its peculiar physiognomy, and even at the beginning of the change of life there is generally something in the appearance which leads me to suspect what is taking place. The physiognomy at such periods indicates debility and suffering, which cannot be accounted for by disease; or there may be the drowsy look, or the dull, stupid astonishment of one seeking to rouse herself so as to answer a question. In several of my worst cerebral cases, there was the same permanent knitting of brows expressive of constant anxiety. An unusual development of hair on the chin and upper lip generally coincides with cessation; so does a sallow complexion; there is also an unusual power of generating caloric, indicated by the habit of throwing off clothing and opening doors and windows even in winter, to the great annoyance of other members of the family. In other respects the appearance of women at the change of life varies according to the type they affect.

Women at the change of life present three types—1st. The Plethoric. 2nd. The Chlorotic. 3rd. The Nervous.

1st. The Plethoric type is unmistakeable, with its turgid tissues, its over-florid countenance in a state of perspiration, its anxious expression, or half-intoxicated appearance. The pulse is often full and bounding, though it may be weak and thready in florid and stout women.

2nd. While in some the change of life increases the proportion of globules in the blood, it diminishes them in others, and thus chlorosis is produced. The Chlorotic type is recognisable by the sallow or semi-chlorotic skin, the weak pulse, the arterial murmurs, and various symptoms of debility. Even on the blanched face, faint flushes will bring out perspiration.

3rd. All women are unusually nervous at the change of

life, but some present a well-marked nervous physiognomy without plethora or chlorosis. The over-anxious look, the brimful eye, the terror-struck expression, as if fearing to see some frightful objects, and the face bedewed with perspiration, are not uncommonly met with.

When the menstrual flow has been for many months absent or irregular, too scanty, or too abundant, in women answering to the above description, I look upon cessation as probable, whatever may be the age, unless the absence of the menstrual flow can be satisfactorily explained by lactation, or pregnancy. In the absence of all other signs, the change of life is to be suspected when, for the first time, towards the forty-fifth year, the menstrual flow comes at irregular intervals.

Cessation may be confounded with I. Chlorosis. II. Inflammation of the womb. III. Uterine Polypi. IV. Uterine Fibrous tumours. V. Uterine Hydatids. VI. Uterine Cancer. VII. Pregnancy. I shall briefly illustrate these assertions by recording the mistakes which have occurred to myself or others.

I. CASE I.—*Chlorosis mistaken for Cessation.*—Anne W., aged thirty-three, and married, had an anæmic hue of countenance. The menstrual flow first came at thirteen; had been regular and without pain until twenty-two, when she married, and had one child at twenty-four. There had been a gradual diminution of the menstrual flow for the previous year, with intense debility, epigastric faintness, and drenching perspirations, and a loud *bruit de souffle* in the carotids. Was it a case of chlorosis in a married woman, or chlorosis occurring at cessation? I inferred the latter from the *gradual* failing of the menstrual flow, and the pertinacity of the flushes and perspirations.

A camphor mixture, a belladonna plaster to the pit of the stomach, and sulphate of iron in pills, cured the patient, and when I saw her again three years afterwards her health was good, but there had been neither pregnancy nor return of the menstrual flow.

II. CASE 2.—*Ulceration of the neck of the womb mistaken for cessation.*—Mrs. J., a delicate-looking lady, with light hair, and grey eyes, was forty-seven. The menstrual flow came at seventeen, without much disturbance. She married at twenty-five, and had three children, the last seven years ago, when her health failed, and the menstrual flow became irregular as to time and quantity, being sometimes very scanty, and at others so abundant as to confine her to the bed or sofa for ten days. There were frequent bearing-down pains, often followed by leucorrhœa. For the last year there had been no menstrual flow, but always a little leucorrhœa, which every month became very abundant. She became nervous, irritable, and was sometimes troubled with flushes and perspirations. She thought the change of life had occurred, and medical advisers confirmed her opinion; but struck with the unusual persistence of leucorrhœa, ever since her last confinement, I advised a further examination, and found the neck of the womb three times its usual size, though little pain was given by pressure. The os uteri was patulous, rough, and extensively ulcerated. This was first treated by the application of the solid nitrate of silver every fourth day, and then by iodine exhibited internally and applied to the surface of the womb, which in six months was restored to a healthy condition, but the menstrual flow did not return. Eight years afterwards, when I again saw this lady, I learnt that a few weeks after I had ceased my attendance, she was taken poorly in the usual way, and that the menstrual flow had appeared monthly ever since. Thus the menstrual flow had been prematurely stopped by the debilitated state of the frame and by the chronic uterine inflammation. That once cured, and the system having gained strength, the ovarian impulse became strong enough to induce the healthy uterine discharge.

Phœbe B. The menstrual flow gradually stopped at thirty-one, after suckling, and as it remained absent for a year, cessation was supposed to have taken place. On examination, however, I found ulceration of the neck of the womb; that being cured by surgical treatment, the menstrual flow went on as usual. In such cases, the error of diagnosis would be prevented by the surgical examination of the patient. In like manner, I have repeatedly known frequent flooding, recurring for years, to be attributed to the change of life, because the patients were about forty-five years of age; whereas, if the practitioners had taken the trouble to make a digital examination they would have found the flooding explained by soft hypertrophy of the cervix or uterine polypus.

III. CASE 3.—*Uterine polypus mistaken for cessation.*— Mrs. P., a thin lady of moderate stature, with dark hair, hazel eyes, and a blanched countenance, consulted me when forty-two years of age. The menstrual flow came at seventeen, and was regular until twenty-two, when she married. She had one child, and menstruation continued regularly for several years; she then became subject to a red discharge, sometimes slight, at others very abundant. She was very weak, had palpitations on the slightest exertion, but no headaches, hysteria, flushes, or perspirations. Her medical man called the case one of premature change of life, but had he made an examination, he would have found a polypus the size of a hen's egg hanging from the cavity of the womb, by a pedicle about as thick as the little finger. I removed the polypus, and the menstrual flow returned, and continued regular for five years afterwards, and I have seen several similar cases, although as a rule, uterine polypi retard cessation.

IV. CASE 4.—*Fibrous tumour of the womb during the change of life supposed to be pregnancy.*—Mary B., aged forty-seven, unmarried, a tall, stout woman, of a sanguine temperament, dark hair, and a florid face. Menstruation first appeared at fifteen, and was regular, and very abundant up to the fortieth year. For the last seven years the flow was always irregular, and had often amounted to a flooding, with intense bearing-down pains. During the same time, she had been subject to frequent flushes and sweats, and to much pseudo-narcotism. Now her eyes look heavy and bewildered, her head feels

giddy and stupid, she is afraid of falling ; there are considerable abdominal enlargement, vesical tenesmus, and morning
sickness. On examination, I found the womb much enlarged,
more so on one side than on the other, owing to a fibrous
tumour of the womb. Subsequently the menses were suppressed for four months, and as the abdomen had increased
in size, she consulted one of my colleagues at the Dispensary,
who, after examination, pronounced her pregnant. Distressed
at an opinion which implied what she knew to be impossible,
she again came to me, and I was able to confirm my former
opinion. The result proved that I was right, for, soon after,
menstruation returned, and continued to do so at irregular
intervals for the following year. In this case, the age of the
patient, the prolonged irregularity of the menstrual flow, the
persistency of pseudo-narcotism, of the flushes and perspirations, made it probable that these symptoms were produced by the change of life.

The sanguineous discharge caused by fibrous tumours and
polypi, and the flooding they determine, are often confounded
with the menstrual discharge and with the flooding of cessa
tion, and this may be correct if the flooding be periodical.
In a case of fibrous tumour, which I shall hereafter relate,
flooding lasted for eight years without intermission ; but
there is no ascertaining whether this was caused by the
change of life.

V. The following case puzzled me exceedingly :—

CASE 5.—*Uterine hydatids supposed to be the change of
life*.—Anne H., a tall, delicate-looking woman, with light
hair and blue eyes, was thirty-nine when she first consulted
me at the Paddington Dispensary, June 17th, 1850. The
menstrual flow first appeared at fifteen, with great pain. She
married at twenty, had six children, and was regular up to
the last year, since which time the flow came at two, three,
five, or eight weeks' interval, with greater pain, and with
nervous fits, in which, though conscious, she would remain
from six to eight hours in a powerless and speechless state
On examination, I found nothing the matter with the womb,

so I imagined the change of life was approaching, and treated her accordingly, but without affording much relief. The menstrual flow stopped for three months, up to which period nothing had caused me to alter my diagnosis; but in December flooding came on, and I have seldom witnessed greater sufferings than she endured for a year, flooding frequently occurring, with intense abdominal pains and hysterical fits. The womb was patulous but not ulcerated, and its body was not much enlarged. I gave opium and ergot of rye in repeated doses for a long time with decided benefit, for on May 10th, 1851, the menstrual flow appeared at the proper time, and without much pain. The abdomen was much swollen. After walking home, on June 23rd, dreadful forcing pains brought away from the vagina about a pint of sticky, rose-coloured water, without smell. This was followed by a sanguineous discharge, lasting for several days. A digital examination was very painful, but I ascertained that the body of the womb was anteverted, much enlarged, and so high that it was difficult to reach the os uteri. I gave opium enemata, which relieved the pains; alum injections were used, but when they stopped the flow, the pains returned, which were only relieved by flooding. To obtain parish relief, it became necessary that the patient should be attended by Mr. Howlett; but I learned that, after very violent pain, half a pint of glutinous liquid was again passed from the vagina on July 5th, after which she continued to lose blood. It sometimes dribbled, at others came as a flooding. When the uterine discharge stopped for two or three days, the pains became excruciating, notwithstanding the exhibition of sedatives inwardly and outwardly; and this state of things continued until I was sent for to see the patient on December 6th, when, after labour pains, she brought away a large mass of hydatids, which well explained the frequent floodings and the patient's protracted sufferings. Ergot of rye induced bearing-down pains, and brought away dark, offensively-smelling blood. Opium relieved the sufferings; the patient recovered, and had another child a year after.

Uterine hydatids is a rare disease, generally occurring

earlier in life, and its diagnosis is often very obscure. The
enlarged womb, the continued flooding without ulceration of
its neck, alternated with the limpid or rose-coloured glutinous
discharge—the best sign of uterine hydatids—was absent in
the early part of this case. The diagnosis of uterine hydatids
being clearly established, it would have been better to have
dilated the cervix and to have brought on the expulsion of
the hydatids, and to thereby save the patient six months'
suffering.

VI. The floodings of cancer are not unfrequently considered indications of the change of life, particularly when
they occur in young women presenting a healthy appearance
—an error which can only be prevented by an examination.

VII. Pregnancy occurring late in life is often taken for
cessation, the more so as, at whatever age, it is not unfrequently accompanied by the flushes, perspirations, and many
of the nervous symptoms which characterize cessation. Under
these circumstances it is better to give a very guarded opinion.
Thus De la Motte relates, that a lady, for fear of having
children, would not marry before fifty-one; and when she
became pregnant, all the symptoms were considered to depend
on the change of life. It much more frequently occurs,
however, that the first symptoms of the change of life are
mistaken for pregnancy. This is sometimes the effect of
imagination; for many women have such a dislike to becoming old that, if married, they would rather persuade
themselves they are with child; and they indulge this persuasion, until, like Harvey's widow, all hope vanishes in
wind. Sometimes, however, the patient's belief in pregnancy
is founded on data which puzzle the faculty. The menstrual
flow stops, the abdomen gradually enlarges, and women, who
have had children, are convinced they feel the moving of the
child, the breasts swell and distil a milky fluid, there is
sickness and impulsive desires; if a flooding takes place, the
symptoms abate, but may again recur, to be again relieved
by flooding, and these floodings of cessation are looked upon

as miscarriages. B. de Boismont relates three cases of this description, and Fodéré two. The following are instances that have occurred in my own practice :—

CASE 6.—*Cessation mistaken for pregnancy.*—Mary C., married, aged forty-five, a tall, delicate-looking woman, with blue eyes and dark hair, thinks herself pregnant, and has been told so by several practitioners. The menstrual flow came at eleven, and was attended by little disturbance both before and after marriage. She married at twenty-seven, but she never conceived. For a year the menstrual flow has been either abundant, with large clots, or very pale, and for the last six months it has been altogether absent, the abdomen gradually enlarging, the breasts becoming swollen and sore, with leucorrhœa and hysterical sensations in the throat. On examination, I found the abdomen enlarged, but without the centrally situated, uniformly round tumour of a pregnant womb. The womb was small, with a virgin orifice. That the patient was not pregnant I felt convinced, and I inferred the change of life from the long continuance of menstrual irregularities, the continued epigastric faintness, the flushes and sweats. My usual treatment relieved the patient; for many successive months she had an abundant vaginal mucous discharge, which was not interfered with. When I last saw the patient she was fifty, and there had been no return of the menstrual flow.

CASE 7.—*Cessation supposed to be pregnancy.*—Sarah B., a tall, thin woman, of a sanguine temperament, says she is thirty-five, but looks forty, applied at the Farringdon Dispensary for a lying-in-letter, on May 7, 1851. The flow first appeared at fourteen, was unattended by pain. She married at eighteen, has had five children, but has never been well since her last confinement two years ago. Hypogastric pains were habitual, the menstrual flow irregular, for the last six times it had the appearance of dirty water, and ceased entirely seven months back, when the abdomen enlarged, and she now feels a fluttering similar to what she felt when pregnant. The breasts are swollen and discharge moisture, and the nipples are surrounded by a dark circle interspersed with a few pseudo-follicles. Notwithstanding these signs of pregnancy, and the patient's assertions, I did not think her pregnant; for there was no solid umbilical tumour, no softening of the neck of the womb; and the irregularities of the menstrual flow, and the appearance it last assumed, coupled with a great liability to flushes and perspirations for the last three months, were my reasons for thinking the change of life had occurred, and I ascertained that I was right upon inquiry several years after.

I have notes of three similar cases.

In the preceding cases, I believe that connexion taking place during the change of life, induced a nervous condition of the ovaries and womb, capable of determining most of the symptoms of pregnancy; and when the abdominal swelling, the nausea, the enlarged breasts persist for months, the supposition of pregnancy might be easily admitted, and the rightful heir to an estate might be defrauded without the possibility of detection, if the woman had already borne children, unless an investigation were made soon after the supposed event. Gooch has noticed another class of cases simulating pregnancy occurring at the change of life. The uterus is torpid, the intestines are flatulent, and the omentum and abdominal parietes have grown very fat, women having then, what Baillie called, "a double chin in the belly." The most notorious blunder of this description was in the case of Joanna Southcott, who, although sixty-four, was thought pregnant by many medical men. The walls of her abdomen were coated with four inches of fat, and the omentum was one lump of fat, and four times the usual size. It is evident that if women marry about the time of cessation, and are fat, and anxious to have children, they may have most of the subjective symptoms of pregnancy—morning sickness, painful breasts, the sensation of something moving in the abdomen, so that it is impossible to affirm the contrary, unless after examination, when the body of the womb will be found of the usual size aud the cervix, hard as in its unimpregnated state. I have kuown womcn led to believe themselves pregnant on account of an cnormous and rapid deposit of fat in the abdomen and breasts only.

Dusourd has met with sevcral cascs of uterinc exfoliation at the change of life, thc membrane bcing passed without flooding, but with great pain ; and this might lead to believe in conception. I have always obscrved thc cxfoliation of the utcrine mucous membrane to occur iu much youngcr womeu, but in thc first case of the kind recorded, Morgagui distinctly states, that the woman contiuucd to exfoliate the mucous membrane of thc womb, up to the cessation of menstruatiou.

In other women, unmarried or married, where no such membrane can bc detcctcd, thc solid coustitucnts of the blood unite, and come away as an ovoid " mole" in the midst of parturient paius. Thcn the womb really cnlarges, and is ablc to causc all the sympathctic disturbance of prcgnancy, as in the following casc :—

CASE 8.—*Cessation and spurious pregnancy mistaken for pregnancy.*—Fivc ycars ago I was consulted by a lady who moved iu good society, was forty-five, unmarried, stout, and of a florid complexiou. The menstrual flow came at fourtecn, had been rcgular and with little disturbance until the prcvious two years, when it bccamc irrcgular, being sometimes very scanty, at othcrs abundant and coutaining largc wellformcd clots coming away with intcnse abdominal pains. Therc had been no menstrual flow for six months previous to my bcing consultcd, and at that timc the abdomcn had gradually attained a size sufficient to attract attention ; hcr appetite was capricious and thcrc was morning sickncss, which circumstances led to ill-natured comments from some malicious acquaintanccs who spread about the report that the lady was in the family way, which report soon reached the cars of a gentleman to whom she had long becn cngaged. On making a careful examination, I found the womb much cnlargcd, but still it was not so largc as it would have bccn

had the patient been six months pregnant. The neck of
the womb was shorter than usual, but not soft as it generally feels in the sixth month of pregnancy. The breasts
were swollen, the nipples surrounded by a darker circle than
habitual; but there were no follicles. Taking these circumstances into consideration, and bearing in mind that
ever since the menstrual flow had been irregular, the patient
had been constantly troubled by flushes and perspirations, I
said that the lady was suffering from the symptoms of
the change of life, complicated by chronic enlargement of
the womb. A course of purgatives, mild tonics, and sedatives somewhat relieved the sickness and nervous symptoms,
but the size of the patient seemed rather to increase than
diminish. Two months after I was first consulted, or eight
months after the menstrual flow had first stopped, I was
sent for in the middle of the night, and found the patient in
pains resembling labour pains, and after these had lasted
several hours, a substance was passed about the size of a
Seville orange. Its external surface was rough, in part
coated with well-organized fibrine, and on being cut open
it was found to contain limpid serum. The walls of this
hæmatic cyst varied in thickness from two to five lines. Its
internal surface had no lining membrane; and microscopical
inspection confirmed the belief in the fibrinous nature of
the cyst. The passing of this singular body was followed
for several weeks by a sero-sanguinolent discharge, with
considerable pain, and by intense mental distress. On
making an examination six weeks after this body was passed,
I found that the womb had returned to its normal size;
the mind was long in recovering tranquillity, for the en-
gagement was broken off, valued friendships had cooled
down to stiff formalities, notwithstanding my emphatic
declaration that the blood cyst passed by the patient could
not be the result of conception. When I last saw the lady
she was forty-nine, and there had been no return of the
menstrual flow.

I believe this kind of spurious pregnancy is most likely

to occur in women who for the first time experience the matrimonial stimulus, at the beginning of the change of life. I have met with several cases of this description; one was a lady who married at forty-seven, after the first indication of the change of life. Another was a lady of forty-five, who after being a widow for twelve years, married again when the menstrual flow had become very irregular. Soon after marriage all the usual symptoms of pregnancy showed themselves, the womb seemed four or five times its usual size, and after remaining so for five months, a mass of half-organized fibrine was passed, and the patient recovered, but the menstrual flow continued irregular. Dusourd mentions the case of a woman, at the change of life, who thought herself pregnant, when on stooping, she felt herself wetted by a serous fluid coming from the womb, then the abdomen subsided as well as the swelling of the breasts. It stands to reason that a uterine or ovarian tumour at the change of life, may be an additional reason for supposing pregnancy.

The signs of pregnancy are, of course, obscure during the first few months, but then, judgment can be deferred. If, at a later period, mistakes are made, it is because all the circumstances of the case are not taken into consideration, or because one is not permitted to ascertain their existence. Thus, in the case of Joanna Southcott, the medical men who believed her pregnant were not allowed to make a vaginal examination; but even without that opportunity Mr. Sims was convinced that the impostor was not pregnant, from the state of her breasts and the absence of any umbilical prominence. In cases simulating pregnancy, the neck of the womb does not feel softened and puffy, the vagina is not livid. The patient states that the abdominal swelling began at the navel, and did not rise from the pubis; the movements felt in the abdomen are accompanied by ill-defined depressions, instead of well-marked elevations. Pressure, cold, and the succussion of the body, do not reproduce the movements as they do when a child is in the womb. The navel is depressed instead of being salient, and there are ill-circumscribed hard-

E

nesses instead of a regular ovoid central mass, but no placental murmur or fœtal tic-tac; although there may be gurgling sounds in the intestines; and the follicular development of the breasts is imperfect. The diagnosis of cessation being thus established, I proceed with my inquiry.

Having ascertained the average date of last menstruation, I have a fixed point whereby to calculate the amount of time before and after which the health of women is more or less unsettled. The length of the dodging time varies, like the prodromata of first menstruation, from a few months to six or seven years. Out of my 500 tabulated cases, there was no dodging time in the 137 women in whom cessation was sudden; 275 thought the dodging time had lasted during the periods indicated in the following table :—

TABLE X.

LENGTH *of Dodging Time.*

TIME.	CASES.	TIME.	CASES.
Months.		Years.	
1	2	2	52
2	3	3	25
3	9	4	9
4	4	5	4
5	5	6	6
6	32	6	5
7	4	8	3
8	11	9	1
9	3	12	1
10	9	14	1
11		18	1
12	60		
18	15		108
			157
	157	Total	265

Average Length of Dodging Time.
Years . . . . 2·2

The average duration of the dodging time would be two years and three months ; but this would give an incorrect idea of its length, for, on cross-questioning patients, I have generally found that for a few months before the menstrual flow became irregular, other changes had taken place, either the quantity had gradually diminished, or was sometimes scanty, at others profuse. Frequently the quality of the discharge had altered, and had been, for months observed to be more serous, of a paler colour, or like brown or green water. I therefore consider that ovarian influence begins to fail in most women, from two to three years before forty-six, or about the forty-fourth year.

There are, likewise, no positive means of measuring the length of time which elapses between the cessation of menstruation and the re-establishment of health, because the resettlement of the constitution is brought about insensibly, and varies extremely in different individuals ; but from the study of 383 cases, I conclude that, in general, when three or four years have elapsed since cessation, women are no longer liable to the floodings, the sweats, or to the distressing nervous symptoms, whether cerebral or ganglionic. I have, however, repeatedly known manifest symptoms of the change of life endure for many years, for twelve years in one instance, and with little abatement. When the cessation of menstruation is to bring into activity the seeds of complaints which have been long dormant in the system, it generally does so in the four years which follow cessation.

Having ascertained the average date of cessation, and taken into account exceptions to the rule, I give the following table, showing the duration of the menstrual function :—

TABLE XI.

DURATION *of the Menstrual Function.*

Number of Years.	Number of Cases. B. de Boismont.	Number of Cases. Dr. Tilt.	Number of Years.	Number of Cases. B. de Boismont.	Number of Cases. Dr. Tilt.
5	1		30	13	36
6	1		31	13	33
8	1		32	9	38
11	1	1	33	9	35
13		1	34	7	49
15		3	35	4	33
16	4	1	36	10	26
17	4	2	37	6	16
18	1	4	38	5	15
19	3	1	39	2	15
20	3	3	40	7	6
21	4	6	41	1	4
22	3	11	42	3	7
23	12	11	43	1	5
24	8	10	44	1	3
25	8	22	45		1
26	11	11	46		1
27	7	25	47		3
28	6	29	48	1	
29	7	33			
	85	174		92 85	326 174
			Total	177	500
Average Duration of Menstrual Function			Yrs.	28·93	31·33

These cases were collected under similar circumstances in Paris and in London; and I account for B. de Boismont having obtained so short an average duration of the menstrual function, from his having operated on so small a number of cases. The average duration of the menstrual function is therefore from thirty to thirty-two years, which is also the average duration of female fecundity, and that of each successive generation, three facts which are interdependent.

The papers in the Medical International Congress of Paris, and Dr. Hannover's excellent opuscule on Menstruation, have enabled me to construct the following table; and I am at a loss to understand how there can be more than four years between the duration of the menstrual function in Denmark and in Norway.

TABLE XII.

Comparative average DURATION *of* MENSTRUATION.

| Countries. | France. | England. | | Den-mark. | Norway. | Russia. |
		London.	Man-chester.			
Number of Cases.	178	500	69	312	391	100
Average Duration.	29·1	31·8	31·2	27·9	32	31
Observers.	Brierre de Bois-mont.	Tilt.	White-head.	Hanno-ver.	Faye and Vogt.	Lieven.

With regard to the duration of menstruation in India we have nothing but the crude beliefs of travellers, and of physicians, who were too lazy or too busy to count.

The annexed table confirms the unproved assertions of B. de Boismont in France, Drs. Guy in London, Frank at Milan, and Dusourd in the south of France, that the duration of menstruation is longest in those who have menstruated earliest, although cessation is sometimes delayed in those who menstruate very late in life, as it has been demonstrated at page 33.

TABLE XIII.

Influence of EARLY AND LATE *Menstruation on the* DURATION OF THE MENSTRUAL FUNCTION.

Years elapsing between 1st and last Menstruation.	Cases of 1st Menstruation at from 10 to 12 Years of Age.	Cases of 1st Menstruation at from 17 to 19 Years of Age.	Years elapsing between 1st and last Menstruation.	Cases of 1st Menstruation at from 10 to 12 Years of Age.	Cases of 1st Menstruation at from 17 to 19 Years of Age.
12		1	31	5	14
13		1	32	5	5
15		1	33	6	6
17		1	34	5	3
18	1		35	5	1
19		1	36	5	
20	2		37	3	
21	3		38	4	1
22	1	3	39	4	2
23	1	9	40	3	
24	1	2	41	1	
25		4	42	3	1
26		1	43	2	
27	1	9	44	1	
28	6	10	45	3	
29	3	4	47	1	
30	1	8			
	20	55		56	33
				20	55
			Total	76	88
Average Duration of Menstrual Function.			Yrs.	34·29	28·78

Thus while the general average duration of the menstrual function is thirty-two years, its average duration increased to thirty-four years and three months in the early menstruated; decreased to twenty years and seven months in the late menstruated, and there is the striking difference of five years and six months between duration of the reproductive life of those who menstruate at an early or late period of youth. This law holds good in Denmark, as will be seen from the following table, extracted from Dr. Hannover's work :—

TABLE XIV.

DURATION OF MENSTRUATION *according to the age at which it begins.*

Dates of 1st Menstruation.	FAIR.			DARK.			TOTALS.		
	No. of Cases.	Dates of Cessation.	Duration of Menstruation.	No. of Cases.	Dates of Cessation.	Duration of Menstruation.	No. of Cases.	Dates of Cessation.	Duration of Menstruation.
12	3	49·67	37·67	2	45·00	33·00	5	47·80	35·80
13	12	45.08	32·08	6	47·50	34·50	18	45·89	32·89
14	23	44·74	30·74	27	45·19	31·19	50	44·98	30·98
15	20	45·30	30·30	14	45·93	30·93	34	45·56	30·56
16	15	44·80	28·80	23	45.35	29·35	38	45·13	29·13
17	17	42·65	25·65	19	43·32	26·32	36	43·00	26·00
18	29	45·28	27·28	20	44·50	26·50	49	44·96	26·96
19	14	44·57	25·57	19	44·95	25·95	33	44·79	25·79
20	12	42·58	22·58	16	47·44	27·44	28	45·36	25·36
21	3	42·00	21·00	7	45·00	24·00	10	44·10	23·10
22	2	45·00	23·00	2	42·00	20·00	4	43·50	21·50
23	3	44·33	21·33	2	42·00	20·00	3	44·33	21·33
24	1	35·00	11·00	3	41·00	17·00	4	39·50	15·50
Total	154	44·51	27·81	158	45·12	28·13	312	44·82	27·97

Is fecundity possible during the change of life? After the forty-sixth year the chance of fecundity suddenly diminishes, becoming less and less every year, but it is possible so long as the menstrual flow appears, however irregularly. I know of two instances in which conception occurred during the change of life. One was a single lady, forty-seven years of age, in whom the menstrual flow had been very irregular for the previous two years, with that general failure of health which so often indicates cessation. The belief that impregnation was impossible at this period led her to permit liberties which were followed by pregnancy and the birth of a child. This case is the more remarkable as connexion only occurred once, seventeen days after a flooding which lasted for ten days, a fact difficult to reconcile with the ovulation theory.

Is fecundity possible after cessation ? This question admits of being answered in the affirmative, because ovulation is not tantamount to menstruation. The ovaries may induce most of the symptoms of menstruation, and they may shed ovules without the womb discharging blood. Women sometimes conceive during lactation without the menstrual flow having returned ; and as, in some very rare cases, conception has taken place before first menstruation, so I believe it possible after cessation in very rare cases. Mr. Pearson, of Staley-bridge, has published the case of a woman who, at the age of forty-seven, was delivered of her tenth child, eighteen months after the cessation of the menstrual flow. In answer to my inquiry, Mr. Pearson informed me that between her two last confinements, three years and four months had elapsed, and that, after suckling the child, she had been regular several months previous to the cessation of the menstrual flow, for which no cause could be detected. This woman suckled her last child and has not menstruated since : forty-seven is no unusual date of childbirth ; but the fact of conception taking place after menstruation had been absent nine months, is very singular. As an instance of the eccentricities which characterize the generative function, I may mention that I know a lady who was married at eighteen ; both herself and her husband enjoyed habitual good health, but conception never took place until the lady was forty-eight, when she bore a child. It would have been the opinion of every medical man that all chance of family had passed, as in another case reported by Schmidt, where a well-formed female, married at nineteen, did not bear a child until she had reached her fiftieth year.

Having ascertained the date of cessation, I shall proceed to enumerate the various modes in which the menstrual flow may cease.

Table XV.

Terminations of the Menstrual Flow in 637 Women.

Modes of Termination.	No. of Cases.
By the gradual diminution of the menstrual flow .	171
By the sudden stoppage of the usual menstrual flow	94
By the sudden stoppage and a terminal flooding .	43
By a terminal flooding	82
By a succession of floodings	56
By alternate copious or scanty menstrual flow . .	36
At irregular intervals, longer than twenty-one days.	99
At irregular intervals, shorter than 21 days . . .	33
At irregular intervals, alternately longer and shorter than 21 days	23
Total	637

When the menstrual flow appeared at irregularly prolonged intervals, it was said to have occurred about every second, third, fourth, fifth, or sixth month, and sometimes at longer intervals, the longest being sixteen months. When it appeared at irregularly contracted periods, it was said to have occurred at about one or two weeks' interval.

The duration of the flow at each menstrual epoch was less and less in the one hundred and seventy-one women in whom there was a gradual diminution of the menstrual flow. When the mode of termination was erratic, the duration of each menstrual period was also irregular. Flooding continued either for a longer or a shorter time than that usually allotted to the menstrual flow; but I have generally observed that, during the first part of the dodging time, the menstrual period was often prolonged to eight, ten, or fifteen days. As this may, however, be caused by some disease of the womb, an examination is necessary. During the latter part of the dodging time the menstrual flow became less and less till it lasted only a day, or was a "mere show." With regard to its quality during this period, it was said to be sometimes blacker than usual, more clotty or sero-sanguino-lent, or to be like cinder-dust and water, or dirty green water, like the latter part of the lochial discharge.

Many physiologists maintain, not only that menstruation is always the effect of ovarian action, but that each menstrual flow is caused by the ovary shedding an ovule, and, under pain of great inconsistency, they must explain in the same way all its varied terminations. In their opinion, the emerging of an ovule from the ovary, causes the floodings of cessation or sero-sanguinolent discharges, no matter whether they occur every week or every six months. I am not obliged to adopt this explanation, for while granting the frequent coincidence of ovulation and menstruation, I have always held 'that the menstrual flow may occur without ovulation, just as ovulation may occur without menstruation. The more I observe, the more I am struck by facts which cannot harmonize with the ovulation theory. I have patients in whom any unusual nervous emotion or over-exertion will bring on the menstrual flow with the usual menstrual symptoms, although they may have only just recovered from this discharge. How can it be supposed that an ovule can be ripened, and the dense ovarian envelope suddenly per-forated, by the fatigue of a dinner party, by hearing dis-agreeable news, or by an altercation with a servant? The laws of ovulation are as yet imperfectly known, but I believe that it proceeds as regularly, inevitably, and uninterruptedly as nutrition; whereas the menstrual function shifts its pe-riodicities, returning about the fourteenth or the twenty-first day after the last epoch, whether it came at a right or a wrong time. This sudden shifting of periodic action is the special attribute of the nervous system, shows the menstrual flow to be impelled by nervous influence, and explains how a strong emotion may repel the menstrual flow or alter the time of its appearance. That sudden emotion should cause the uterine surface to perspire is only a repetition of the well-known effects of emotion on other parts.

Obesity.—The evident effect of the change of life is to re-model the female frame, so that health may be made consistent with the absence of an habitual drain. An unusual intensity of that force which presides over nutrition is shown by its being able in many women to use up the retained blood, so as to strengthen the tissues of the frame, in the same way that it improves their outward appearance. In others, the nutritive force stores up fat for future emergencies, and when cessation has actually taken place, the accumulation of fat is similar to that which occurs in animals from whom the ovaries have been removed in early life. I have seen the sudden growth of fat coincide with great improvement in health, but Dr. G. Bedford goes too far when he states "that those who became fat-bellied were not troubled with nervous symptoms." If the vital force is not able to turn to good account the superabundant blood, and cannot maintain an efficient control over the vaso-motor nerves, a host of nervous symptoms arise and the balance of circulation is disturbed, so as to lead to erratic fluctuations, mis-directions of blood, and to congestions or discharges from various organs. This occurs generally during the first part of the change of life, when women rather decrease in size than otherwise, even if they were to grow fat in the latter part. Thus out of 383 women in whom there had been no menstrual flow for five years, 121 had grown stouter than before, 71 had retained the same size, and 90 had become thinner; in other words, only one-fourth of the 383 had become thinner than usual five years after cessation. Wishing to know whether continued illness during the change of life had any effect on the greater or less tendency to become stout, I divided the 383 women into three classes, according as they had suffered much, little, or not at all, and I found that the proportions of the stout and thin were about the same, and that if those who suffered considerably during the dodging time had then become thinner, nutrition became unusually active in the latter part of the change of life. Three out of the 383 became suddenly fat at cessation, but this *embonpoint* was no indication of strength. Although the mammary glands become atrophied at the change of life, they often seem to retain their former size from the deposit of fat in their vicinity, and thus become pendulous. Fothergill has observed that women are more likely to grow stout in whom menstruation has suddenly ceased before the usual time, unless it be caused by some internal complaint. I have known this to occur several times, as in the following instance, which may help the diagnosis of early cessation :—

CASE 8.—*Sudden embonpoint a sign of cessation.*—Eliza L., aged thirty-two. The menstrual flow appeared at nine, and came regularly with a moderate amount of headache and pseudo-narcotism. She married at twenty-seven, but never conceived, and was regular until the menstrual flow suddenly ceased at thirty, after a fit in which she was unconscious for many hours. When a child at school, she used to faint away, and was with difficulty brought to her senses. She frequently fainted before marriage, but since then no fainting fits had occurred until two years ago, when, after unusual exertion, she remained unconscious for five hours. A year ago she had another fainting fit, and for the last two years she has been much troubled with giddiness, headache, and pseudo-narcotism. She now complains of not being able to see, " as if from a skin before the eyes." She is habitually troubled with epigastric pain, or uneasy sensations at the pit of the stomach, and she has grown immensely stout. Ten years afterwards I ascertained that there had been no return of menstruation.

As for thirty-two years it had been habitual for woman to lose about 3 oz. of blood every month, so it would have been indeed singular, if there did not exist some well-contrived compensating discharges acting as wastegates to protect the system, until health could be permanently re-established by striking new balances in the allotment of blood to the various parts. The compensating agencies may be thus classed :—

I. A larger consumption of carbon by the lungs.

II. An unusual amount of urinary deposits.

III. Increased perspiration.

IV. Abundant mucous flows.

V. Hæmorrhages from various organs.

Some of these compensating actions proceed permanently, as from the surface of the lungs and skin; others occur irregularly; but in a certain number of cases the compensating action recurs periodically, assuming the monthly type, the type of the function which is falling into disuse.

TABLE XVI.

Monthly Occurrences after Cessation in 53 out of 500 Women.

Nature of the Occurrence.	No. of Cases.
Lumbo-abdominal pains	15
Leucorrhœa	12
Headache and pseudo-narcotism . . .	7
Diarrhœa	5
Entorrhagia	2
Bleeding piles	1
Hysterical symptoms	2
Hysterical oppression, or asthma . . .	2
Great depression of strength	1
Sweats	2
Dyspepsia	1
Stomatitis	1
Swelled gums	1
Swelled legs for 3 days	1
Total	53

I. RELIEF AFFORDED BY RESPIRATION.—The balance between the respiratory and the menstrual function is forcibly shown by the fact that, at all periods of life, whenever menstruation is diminished or suppressed, more carbon is consumed per hour by the lungs, and less when the menstrual flow has been re-established. Andral and Gavarret found that, in both sexes, from eight years of age to puberty, there was an augmentation in the quantity of carbonic acid gas excreted by the lungs; that whenever menstruation takes place in women, they still continue to excrete the same quantity of carbonic acid as before; while in man, at the same age, the quantity goes on increasing until he has reached the middle period of life, after which he is said to secrete less and less. From experiments made on twenty-one women, it appears that during the whole of the time comprised between the first and last menstruation, the strongest and healthiest only excreted per hour a quantity of carbonic acid representing grammes 6·4, or the same quantity as before puberty, while men excreted grammes 7·8 per hour before their fifteenth year, and 11·3 per hour from fifteen to forty years of age. After cessation of the flow, the lungs more largely excrete carbonic acid; for in five women aged from thirty-eight to forty-nine, and who had ceased to menstruate, the quantity of carbon consumed per hour rose from grammes 6·4 to 8·4. Women likewise excrete more carbonic acid from the lungs during pregnancy and amenorrhœa, which may help to explain how the seeds of consumption may be developed under these conditions.

II. RELIEF OBTAINED BY THE INCREASED DEPOSITS IN THE URINE.—The majority of women notice that, during the dodging time the urine, instead of being transparent, is unusually loaded with sediments, and this persists for weeks, and often recurs during that period. These sediments consist of lithates and phosphates, and, as in febrile affections, they indicate an effort of nature to relieve the system, and point to the utility of alkalies.

III. RELIEF OBTAINED BY INCREASED SECRETIONS FROM THE SKIN.—The importance of the relief afforded by the skin may be gathered from the subjoined table :—

TABLE XVII.

Cutaneous Exhalations in 300 out of 500 Women.

	No. of Cases.
Perspirations attendant on flushes	201
Perspirations occurring monthly	2
Sweats	84
Cold perspirations	13
Total	300

With the exception of the few cases in which the perspirations were cold and clammy, this exhalation is associated with an increased production of heat, and with that irregular distribution of it which is called "flushes." These occurred to 244 women out of 500; fourteen others were troubled with "dry flushes." The flushes determine the perspirations, and, as they constitute the most important and habitual safety-valve of the system at the change of life, it is worth while studying them.

It must have struck many, that at the change of life most women have the power of generating a more than usual amount of heat; they often want less clothing, and even in winter leave their doors and windows wide open. Sometimes, however, instead of being regularly distributed, this caloric bursts forth as flushes, women feel as if something started from the epigastric region, spreading over the chest, and then over the face, which becomes suffused and hot. I have heard women compare their sensations to burning steam rising from the pit of the stomach. These flushes may be considered as cases of pathological blushing, depending on partial paralysis of the vaso-motor nerves. They sometimes come after a chill, or a momentary sensation of shivering, or after sinking and faintness at the pit of the stomach, but oftener without these sensations. If the flushes do not terminate in a gentle moisture of the skin, women call them "dry heats." The flushes may last two or three minutes,

but it will often take a quarter of an hour to carry off the
effects of each wave of heat that is wafted to the surface.
The number of flushes occurring in the course of the day
varies extremely. Their spontaneous repetition, five or six
times an hour, either by day or night, is not uncommon to
women at the dodging time. Some months after cessation
they frequently occur only seven or eight times in the day.
This may be the case for years after cessation, and even in
extreme old age, under the influence of worry or ill-health,
without however having the intensity which renders them
so distressing at the change of life. The recurrence of
flushes so late in life is not to be wondered at, for women
can blush at sixty or seventy years of age.

The face and chest generally suffer most from flushes, but
the whole of the skin may be affected, and the hands and
nails may feel like fire, the pulse being often weak and slow.
Robust women, of a sanguine temperament, are more troubled
with flushes, confirming Sir J. Ross's assertion, that the
sanguine temperament has a peculiar power of generating
heat, denied to the pale and sallow. All molecular action
generates heat, and while this action incessantly proceeds in
living bodies they generate heat as permanent and continuous
as their nutrition, beginning with life to cease only with it.
The quantity and quality of this heat varies with the quantity
and quality of the blood, which is the fuel of the animal
combustion, proceeding in the breadth and depth of our
tissues, so that if the blood is in a healthy condition the
heat is physiological, if on the contrary, it is the heat of
fever; but how is heat collected? and what are the laws of
its distribution? At times it will burst forth in fitful paroxysms; at times, as in intermittent fevers, these ebullitions
of heat follow a rhythmical march, and depend on diseased
action of the vaso-motor nerves, which hold the blood-
vessels in their web-like grasp. One feels cold before dinner;
a few mouthfuls of solid food are taken and sensations of
warmth are produced, not to be explained by the assimilation
of yet undigested food; that is *nervous* heat. When the food

turns into chyle, passes into the blood and becomes the pabulum of the chemical actions of nutrition, there is a marked increase in the amount of *chemical* heat. Blushing exemplifies nervous heat. The mental portion of our being first receives an impression, and instantaneously communicates it to the emotional; then follows the sudden reaction on the epigastric ganglia ; a momentary loss of power takes place there, during which the skin is somewhat paler and colder than usual, there may be an imperceptible sigh, a glow is felt at the precordial region, a sudden something seems to rush forth from the epigastrium, swift as lightning, and then wave after wave of blood is poured in burning streams to the whole surface, or only to the neck and face, which appear to blaze with living blood. The heat passes off, and the blood retreats from the capillaries, resuming its slower course, unless the emotional feelings become once more aroused. Thus, in the physiological phenomena of blushing, heat is evolved by the reaction of emotion on the voluminous ganglia and ganglionic nerves situated at the pit of the stomach ; what this nervous centre does under the influence of emotion it also does *spontaneously*, causing the heats or flushes to which women are subject at the menstrual periods, during pregnancy, and lactation, but chiefly at the change of life ; in fact, whenever the functions of the ganglionic nervous system are disturbed. This confirms the assertions of the oldest physiologists, for Hippocrates noted shivering and an unusual development of caloric to be a sign of conception, Galen considered the reproductive organs to be a source of caloric, and Brongniart discovered that even in plants, reproduction is attended by an appreciable increase of their usual temperature.

It was well known that any great depression of nervous power paralysed the vaso-motor nerves of the skin, causing it to sweat, but we owe to Claude Bernard the establishment of the fact that while the section of the sensory and motor nerves is not followed by any increase of heat, on the contrary, the division of the nerve by which the superior and inferior

cervical ganglia communicate, by preventing the influence of
the great sympathetic nerve from proceeding to its central
ganglia, developed great heat in the capillaries above the
point of section of the ganglionic nerve.   Dr. H. Jones found
that the division of a sympathetic nerve causes the blood-
vessels within its range of distribution to dilate, the pulsation
of the arteries to become more energetic, and the temperature
to rise as much as 10° and 15°.   When the sympathetic
nerve is divided on one side of a horse's neck, that side of
the face and neck becomes bathed in sweat.   Something
similar occurs at the change of life; the demise of ovarian
activity half paralyses the ganglionic system, and the blood-
vessels, thus leading to a greater development of heat, and
to its irregular emission from the surface.   Flushing, like a
fit of ague, has a period of concentration, a hot stage, and
one of perspiration; but as in ague there is often no first
stage, so flushing frequently comes first, and often, without
any previous congestion of the capillaries, perspiration is seen
continually oozing out of the skin, where it stays until it is
wiped away or rolls off.   Why continued perspirations are
so frequently met with at the change of life, and how they
preserve women from worse evils, cannot be answered without
some inquiry into the physiology of the skin.

With regard to the extent of the surface operated upon by
nature, the entire surface of the body being estimated as
over ten square feet, it has been calculated that there are
70,000,000 pores through which perspiration flows, besides
the openings by which the sebaceous glands emit their lu-
bricating products.   Lavoisier estimated that, in twenty-four
hours, twenty ounces of perspiration were lost by the skin.
Dr. Southwood Smith has shown that a robust man, working
hard in an intense heat, may lose five pounds of his weight
in an hour, so that the emaciation of those who perspire
much is to be understood, as well as the *embonpoint* of those
who perspire little.

Long ago Sanctorius established that "Insensible perspiration alone discharges much more than all the other

evacuations put together." Valentin found that his average hourly quantity of perspiration amounted to 463·3 grains, and that by walking up and down hill, to sweat copiously, more than four times this quantity was excreted, or 2048·8 grains. The average proportion of water in the human blood is 78 per cent., so that sweating gets rid of its redundant portion; and with regard to the other components of perspiration, Dr. Favre, of Paris, obtained from a healthy man, 14 litres or 28 lbs. of perspiration, which he found to be thus composed :—

TABLE XVIII.

	Pour 14 Litres.	Pour 10,000 Grammes.
	Grammes.	Grammes.
Chlorure de sodium . .	31,327	22,305
Chlorure de potassium .	3,412	2,437
Sulfates alcalins . . .	0,161	0,115
Phosphates alcalins . .	Trace	Trace
Albuminates alcalins . .	0,070	0,050
Phosphates alcalino-terrenes	Trace	Trace
Débris d'epiderme . . .	Trace	Trace
Lactates alcalins . . .	4,440	3,171
Sudorates alcalins . .	21,873	15,623
Urée	0,599	0,428
Matières grasses . . .	0,191	0,177
Eau	13,938,027	9,955,737

Solubles dans l'eau. / *Solubles dans l'eau acidulée.*

Dr. Favre draws the following conclusions from his researches :—

1. That the materials of perspiration are almost entirely soluble in water.

2. That chloride of sodium constitutes the chief solid constituent of perspiration, the alkaline sulphates and phosphates being in very small proportion. In this respect, perspiration differs much from urine, as will be seen from the following analysis of equal quantities of each :—

	Perspiration 14 litres.		Urine 14 litres.
Chlorides .	grammes 34,639	grammes	57,018
Sulphates .	„ 0,160	„	21,709
Phosphates .	„ Trace	„	5,381

3. Lactic acid, combined with alkalies, exists in the perspiration. It would be worth while ascertaining whether its quantity be augmented or not in the acid perspirations of lying-in women, in the sweats of over-lactation, or in those of the change of life.

4. Sudoric acid combined with alkalies, a hitherto un-noticed component of perspiration, is one of its principal elements. The composition of this acid is somewhat similar to that of urea, which is not normally found in perspiration ; but when perspiration is stopped, as in the experiments of Seguin and Anselmino, the kidneys may transform sudoric acid into uric, and excrete it as urate of ammonia.

5. The existence of urea in perspiration completes the analogy of this fluid with urine.

6. Perspiration taken at different times from the same person did not vary in composition, but when the fluid taken during the first half-hour of the experiment is compared with that taken during the second or third half-hour, a larger quantity of mineral salts is found to be expelled during the latter periods. The proportion of water to the other com-ponents of perspiration does not seem to vary.

At the change of life the quantity of perspiration is often evidently increased, but we do not know how far its various components may be altered in amount.

As the skin is the most easily moved of all the safety-valves of the system, it is most frequently influenced by the change of life ; while the small proportion of solid removed from the blood by sweat, explains why this forms an ineffectual crisis, the necessity for which is ever recurring, unless other changes have been effected. The critical nature of the per-spirations occurring at the change of life, is shown by their appearance after sudden suppression of the menstrual flow, after diarrhœa or leucorrhœa. Thus in Eliza C., aged sixty, the menstrual flow suddenly ceased at forty-three, after getting her feet wet; and the perspirations were often so intense that the bedclothes could be wrung. Mary M., aged fifty-five; the menstrual flow came at nineteen and left at thirty by a flooding; she has ever since been subject to flushes and perspirations, by which her health has been preserved. After cessation, the perspirations generally diminish, though I have noticed drenching ones to return five years after, in consequence of great anxiety, and twenty-nine years after, in a strong healthy old woman of sixty-seven. The perspirations do not always cover the whole body, but are limited to certain portions. Two patients always wear flannel at the pit of the stomach, because they perspire so much there, and there only, that their clothes become saturated. Like Tissot, I have seen gentle perspirations converted into sweats for a few days every month.

IV. RELIEF AFFORDED BY MUCOUS FLOWS.—The following table shows the frequency of mucous flows at the change of life :—

TABLE XIX.

Frequency of Mucous Discharges amongst 500 *Women at the Change of Life.*

Nature of Mucous Flow.	No. of Cases.
Leucorrhœa at irregular intervals . .	146
Monthly leucorrhœa	12
Repeated vomiting of mucus	31
Water-brash	5
Frequent diarrhœa	45
Monthly diarrhœa	5
Total	244

These mucous discharges must be considered as taking the place of the menstrual flow, for they were usually preceded by the same symptoms, particularly when they came periodically, a fact already noticed by B. de Boismont. The critical nature of these discharges shows that they may require to be restrained, though not stopped. Of 260 women in whom the menstrual function had ceased, 143 had never been subject to leucorrhœa; of the remaining 117—

The vaginal secretion was increased at cessation in .   77 cases.
It was diminished in . . . .   24 „
It remained stationary in . . . .   16 „

With regard to periodical leucorrhœa. In one case this occurred regularly every month, for a year, for eighteen months, in another, in several for two years, and in one for seven years. Most of these cases occurred to women in whom the menstrual flow had suddenly ceased.

V. RELIEF AFFORDED BY HÆMORRHAGES.—The extent to which women are relieved by hæmorrhages is shown by the following table :—

TABLE XX.

Comparative frequency of Hæmorrhages at the Change of Life amongst 500 Women.

Varieties of Hæmorrhage.	No. of Cases.
Terminal menorrhagia	82
Successive attacks of menorrhagia . .	56
Bleeding piles	24
Entorrhagia	20
Epistaxis	12
Hæmoptysis.	8
Cerebral hæmorrhage and apoplexy . .	6
Hæmatemesis	4
Hæmaturia , ·	1
Bursting of varicose veins	3
Bleeding from external auditory canal ,	4
Cutaneous ecchymosis	3
Bleeding from a temporal vein	1
Total	208

Vicarious menstruation is rare at the change of life, but I have known a bloody discharge from the nipples to occur every month, for five years after the menstrual flow had ceased; and in another case, a sero-sanguinolent discharge came from the nipples at irregular periods, during the two years that followed the ménopause. Dr. Semple has recorded the case of a woman, aged eighty, who, every other month, bled from the nose, or menstruated as in youth. These various hæmorrhages are more or less successfully critical; the floodings at the change of life prevent more serious illness, and afford time for the gradual readjustment of the system. Successive floodings occur as often as the frame is overloaded by blood, returning at longer intervals and to a less amount.

Having thus concluded my study of the change of life, in a physiological point of view, I shall be able to point out briefly in the following chapters the special bearings of the change of life on general pathology.

PART II.

GENERAL PATHOLOGY

OF

THE CHANGE OF LIFE.

CHAPTER III.

A REFERENCE to the works written on diseases of the change of life by Gardanne and Dr. Menville would show the reader, that they consider as diseases of the critical time all which may afflict a woman after her forty-sixth year. This is perfectly absurd, for at any period of life we are liable to many diseases that are not caused by it, and as women grow old they are liable to the diseases of old age, for both sexes suffer alike from that wear and tear of life which impairs the structure of the heart and arteries, and indeed of most organs. I call "diseases of the change of life," such as occur for the first time, or recur with great aggravation during that period, variable as it is in each individual, but generally comprising the three years previous to, and the five years subsequent to, cessation.

The affections of the change of life sometimes come on before any marked irregularities in the menstrual flow: thus a tall, portly lady, aged fifty, is still regular, but for the last few months she has suffered from burning flushes, and her hands often become scarlet to the tips of her fingers. Some have merely flushes and slight perspirations for about three to five years after cessation, at the end of which time severe symptoms occur, and I attribute them to the change of life, because they are exactly similar to symptoms frequently supervening as the immediate result of the ménopause. Sometimes diseases of the change of life occur from the sudden interruption of the heats and flushes which had hitherto proceeded uninterruptedly ; at other times the skin continues to act as the safety-valve of the system, until the health breaks down under the aggression of some mental or physical shock. In some, the advent of disease frequently coincides with an imperfect menstrual nisus, indicated by abdominal pains and a menstrual flow, which, although very slight in quantity, and recurring three or five years after cessation, clearly indicates an effort of menstruation. Gout, rheumatism, or nervous affections, first occurring at the change of life, may continue, more or less till the end of life.

It is well known that, at some time or other of life, each organic apparatus becomes a normal or abnormal focus of action for the general circulation; for instance, the genital organs at puberty and cessation. Each well-marked period has its own particular stock of pathological calamities, and if the diseases common to one epoch do not cease at the beginning of the next, they generally continue during the whole period, or in other words, critical times are *perfect*, leading to the renewal of health, or *imperfect* and determining a succession of discharges which should not be suddenly suppressed. The change of life not only determines some diseases for the first time, but finding also the germs of others, such as gout, cancer, &c., it gives them an additional impulse. In the absence of trustworthy information, I have endeavoured to obtain a correct bead-roll of the infirmities entailed by the change of life, and have set down very minutely whatever could be detected of morbid in 500 women who sought advice for diseases of the change of life. The following table, therefore, gives the comparative degree of frequency of morbid symptoms at the change of life, and it will show the reader, at a glance, what are the actual morbid liabilities of this epoch.

Table XXI.

Relative frequency of Morbid Liabilities at the Change of Life in 500 Women.

Nervous irritability	459	Inflammation, or ulceration of the neck of the womb	11
Flushes	287	Deafness	11
Pseudo-narcotism	277	Follicular inflammation of the vulva	10
Dorsal pain	226		
Gangliopathy and faintness	220	Difficulty and pain in passing urine	9
Headache	208		
Abdominal pain	205	Hysterical asthma	8
Perspirations	201	Intense hysterical flatulence	8
Leucorrhœa	146	Hæmoptysis	8
Hysterical state	146	Monthly headaches	7
Flooding once or repeatedly	138	Rheumatism of joints	7
Sick headache	92	Ulcerated leg	7
Sweats	84	Diarrhœa, from three to six times daily for years	7
Piles, not bleeding	62		
Biliousness, obstinate	55	Intercostal neuralgia	6
Gangliopathy, or strange epigastric sensations	49	Paraplegia	6
Phosphatic and lithic urine	49	Jaundice	6
Diarrhœa, irregular	45	Apoplexy and hemiplegia	6
Debility, intense and long continued	41	Supposed pregnancy	6
Chloro-anæmia	40	Colics, habitual for years	6
Dyspepsia	37	Nettlerash	5
Remittent menstruation	33	Ovarian pain, constant	5
Vomiting, repeated	31	Waterbrash	5
Abdomen much swollen	26	Epilepsy, increased	5
Fainting away repeatedly	25	Prolapsus of womb	5
Piles, bleeding	24	Cancer of womb	4
Constipation, obstinate	23	Inflammation of vagina	4
Entorrhagia	20	Bronchitis	4
Œdematous legs	19	Inability to retain urine	4
Undetermined cutaneous eruptions	18	Laughing and crying fits, only since cessation	4
Globus hystericus, only since cessation	17	Labia, repeated inflammation of	4
Regular monthly pains without menstrual flow	16	Numbness, pricking and loss of sensation in arms and hands	4
Insanity	16	Uterine fibrous tumours	4
Dry flushes	14	Hæmorrhage from ears, increased	4
Mammary irritation and swelling	14	Prolonged fits of unconsciousness	4
Perspirations, cold	13	Vomiting of blood	4
Leucorrhœa, monthly	12	Polypus of womb	4
Epistaxis	12	Erysipelas	4
		Sciatica	4

Epilepsy, brought on	3	Abscess in fingers	2
Delirium	3	,, neck	2
Pus in motions	3	Gout	2
Cutaneous ecchymosis . . .	3	Peeling of nails	2
Prurigo	3	Erectile tumour of the meatus	
Puffed face	3	urinarius	2
Inflammation of legs, with great		Brow ague increased	2
distension of veins	3	,, brought on	1
Eczema	3	Hæmaturia	1
Ovarian tumours	3	Monthly failure of strength . .	1
Aching under nails	3	,, hysteria	1
Inflamed rectum	3	Hysterical paralysis of arms .	1
Hysterical fits, only since cessa-		Falling-off of all the nails . .	1
tion	3	Peritonitis	1
Hemicrania	3	Chronic otorrhœa	1
Consumption aggravated . . .	3	Inflamed eyes	1
Bleeding from varicose veins of		Toothache	1
leg—in one case three times .	3	Morbus cordis	1
Eczema increased	2	Shingles	1
Monthly blood in motions for		Herpes circinatus	1
six months	2	Mammary cancer	1
Boils in seat	2	Varicose veins brought on . .	1
Aortic pulsation	2	,, aggravated . .	1
Legs burning and painful . .	2	Bleeding from frontal vein . .	1
Intensely hot hands	2	Limpid secretion from breast,	
Abscess in arm-pits	2	for four years	1

In my first edition, I stated that the real diseases of the
change of life were the morbid exaggeration of some phenomena natural to cessation. Thus, as flooding occurs as a
natural phenomenon at cessation, it may be attended by
circumstances which may constitute it a disease. As increased perspiration is a natural phenomenon of cessation,
so it may occur to a morbid amount under the form of
continued sweating. The mild forms of gangliopathy are frequent accompaniments of cessation; therefore, the severer
forms of the complaint may be expected. Cessation is almost
always attended by slight and varied cerebral disturbance;
therefore, severe nervous affections may frequently be expected. The foregoing table admirably confirms the views
I have long entertained, and it shows that the *real* diseases
of the change of life are the first twenty-five on the above
list : they were the *common* complaints of this critical period,
whose occurrence has been so frequent that they may be

predicted; prevision being the only criterion of science, since it permits prevention. After the first twenty-five, the frequency of other diseases falls below five per cent., for they either form special varieties of the more common complaints already enumerated, or, like cancer and gout, they may occur at other times besides the change of life. Sometimes, also, by the singularity of their nature and by the rarity of their occurrence, they are evidently shown to be personal contingencies, indicating how the impetus of morbid action was most felt by the weakest organ. The table likewise shows that the diseases to be really feared at the change of life are not so much cancer and other organic diseases of the ovaries and womb, with which women continually frighten themselves, as the neuralgic affections of the cerebro-spinal and ganglionic systems, and that gastro-intestinal affections are more to be feared than floodings.

From the foregoing table it is clear that functional diseases of the nervous system are frequently met with. The table does not so clearly show the frequent occurrence of debility, it is marked as the only disease in forty-one cases, but in a less degree it accompanies chlorosis in forty cases, and the two hundred and twenty cases of epigastric faintness, and was indeed rarely absent in all cases; so I conclude that diseases of cessation are characterized by debility. This loss of strength causes diseases of cessation to be generally chronic, and contrary to what occurs in childhood, pathological movements slacken pace. Malignant disease that at the age of thirty would have killed the patient in three years, may last longer. There is a chance of ovarian tumours remaining in abeyance which at twenty-five would have quickly galloped through their course. Ulceration of the womb takes longer to cure. The eschar made by nitrate of silver may take twelve days instead of four before it comes away from the womb, and the eschar made by the acid nitrate of mercury three weeks instead of one; hence the *prognosis* of diseases of cessation should be guarded, as in chronic disease.

Prognosis.—The prognosis of diseases at the ménopause is

in general satisfactory, unless it be question of malignant diseases, very voluminous fibroid or ovarian tumours, and when the crisis so paralyses the ganglionic centres that the patients cannot recover strength, and continue for years to lead a life of which every act is stamped with debility, not to be accounted for by organic disease.

Before sketching my theory of diseases at the ménopause I shall briefly pass in review their causes, a correct estimation of which is indispensable to understand the pathology of this critical period, and these causes are :—

I. Weakness of constitution.—II. Temperament.—III. Constitutional disorders.—IV. Uterine affections.—V. Unusual suffering at puberty and at menstrual periods.—VI. Sudden cessation.—VII. Disuse or abuse of the reproductive organs.—VIII. Social position.

I. WEAKNESS OF CONSTITUTION.—The best way to avoid the dangers of this critical time is to meet its approach with a healthy constitution. A marked want of strength prevents the regular succession of the vital phenomena by which all critical periods are carried through; and as the change of life is marked by debility when this is grafted on constitutional weakness, loss of power will be of long duration, sometimes with a tendency to faint, to chlorosis, to erratic nervous disorders, and all complaints remain chronic because there is not stamina enough to carry them through their stages.

II. TEMPERAMENT.—I have stated that women of decidedly lymphatic temperament often derived marked benefit from the change of life. Their colder nature is less liable to be disturbed by bilious and nervous disorders, while the blood that was wont to be eliminated is turned to the improvement of the frame. Women of a sanguine temperament are naturally most liable to plethora, and therefore to floodings, hæmorrhages, and apoplexy; but if they are most subject to dangerous diseases, they are most amenable to treatment, and speedily improve by active measures when judiciously applied. The preceding table shows a decided tendency of women to bilious and gastro-intestinal affections at the change of life; these affections being severe in women of a bilious temperament, as Gardanne remarked. Calomel and blue pill may long be given for an habitually torpid liver with little benefit, unless combined with the free use of alkalies, or the taking of mineral waters. Women are more liable to insanity, if the nervous be associated with the bilious temperament, as in some of my worst cases. Women of a nervous temperament suffer much more than others, particularly during the dodging-time; they are as liable to flooding as the plethoric, without being so amenable to treatment. The floodings of the nervous are the result of nervous erethism; in them simple things bring on inordinate reactions, and the common functions of life are performed with eccentricities without number. Such women seldom recover health unless they gain flesh.

III. CONSTITUTIONAL DISEASES.—It is evident that all constitutional affections will be increased by the change of life, and that an impulse will be given to cancer, gout, or consumption. These complaints, in their turn, aggravate the ordinary symptoms of the ménopause.

IV. UTERINE DISEASE.—For a long time after cessation, the womb is congested every month, as is proved by the frequency of monthly abdominal pains and of leucorrhœa. This state of congestion, without the usual adequate relief, is eminently calculated to give rise to congestion and irritability of the womb, and I believe inflammatory affections of the neck of the womb, during the dodging time, to occur more frequently than is represented in the last table. Should cancer impend, morbid irritability and congestion of the womb at the change of life increases the liability to that disease.

V. UNUSUAL SUFFERING AT PUBERTY AND MENSTRUAL ECCENTRICITIES.—The following table shows at a glance the morbid liabilities of puberty compared with those of the change of life; explains to a certain extent the occurrence of other complaints at those periods only, and how the perils of cessation may be inferred and measured by those which have attended puberty:—

TABLE XXII.

Comparative Morbid Liabilities at Puberty and at Cessation.

DISEASES.	PUBERTY.	CESSATION.
Headache	Frequent	Frequent.
Sick headache . .	Frequent	Frequent.
Pseudo-narcotism .	Frequent	Frequent.
Minor forms of hysteria	Frequent	Frequent.
Hysterical attacks .	Rare	Rare.
Delirium	Very rare	Very rare.
Irregular temper and mild forms of moral insanity. .	Frequent	Frequent.
Epilepsy	Frequent	Very rare.
Gangliopathy. . .	Mild forms frequent .	{ Severe and mild forms frequent.
Fainting	Common	Uncommon.
Chlorosis	Very frequent . . .	Not uncommon.
Lumbo - abdominal neuralgia . . .	Very frequent . . .	Very frequent.
Neuralgia of limbs .	Very frequent . . .	Not frequent.
Mammary neuralgia	Not uncommon . .	Not uncommon.
Flooding	Rare	Common.
Epistaxis	Frequent	Uncommon.
Piles	Unobserved . . .	Frequent.
Leucorrhœa . . .	Frequent	Frequent.
Cutaneous eruptions	Frequent	Rare.
Gout, its irregular forms	Sometimes observed .	Not unfrequent.
Rickets.	Not uncommon . .	{ Not noticed by me, but by B. de Boismont, and Gendrin.
Spinal deviations .	Very frequent . . .	Very rare.

Other diseases which may have preceded first menstruation may likewise be expected to precede its cessation; thus Alibert informed me that he had observed some cutaneous eruptions to appear twice only in life—viz.: before the first appearance of the monthly flow, and at its cessation. B. de Boismont, and others, likewise notice the appearance of hysteria and epilepsy before these two important epochs, the patient's life having been free from these diseases during the intervening period. Sir H. Marsh frequently noticed that women, in whom the establishment of puberty had been preceded by repeated epistaxis, experienced the same accident as a prominent symptom of cessation. In my own practice, I have several times seen puberty and cessation preceded and attended by an abundant eruption of boils, by long-continued otorrhœa, frequently by continued diarrhœa, and still oftener by a great amount of pseudo-narcotism and hysteria in cases where there was little or none, while the function was regular, during child-bearing, or lactation.

Spinal deformity slightly developed at puberty may become very marked at cessation, as in a case given by B. de Boismont, where a lady, comparatively straight, became deformed at the change of life, and I have published a similar case in which the pain attending internal metritis was so great that I could never get my patient to lie down flat, and she always slept in a crouching posture. The fact of the menstrual periods having often been unusually painful, is a reason to fear that the change of life will be fraught with unusual suffering, and the same applies to those in whom the menstrual flow has been very irregular, and liable to deviations. Thus B. de Boismont relates of two women, in whom the critical discharge took place from the mouth, even after they had borne children; that at the change of life, one became rickety, the other had ascites. In some, the liability to disease at this period, might be referred to the predominance of a nervous temperament; in others, it depends on hidden peculiarities, which render the performance of the various acts of the reproductive function calamitous through life.

VI. SUDDEN STOPPAGE OF THE MENSTRUAL FLOW.—It was logical to suppose that a drain, lasting for thirty-two years, could not suddenly cease without causing serious illness; and Meissner, like many others, attributes to this cause the most disastrous effects; incorrectly, however, for I have not noted sudden cessation as frequent in those who suffered most at the change of life. Thus, in 383 women in whom the menstrual flow had ceased, 223 had suffered much, 121 little, and thirty-nine not at all. Sudden cessation occurred in each class in the following proportion :—Fifty-five out of the 223 who suffered most, thirty-one out of the 121 who suffered little, and twenty-three out of the thirty-nine cases who did not suffer ; so that while sudden cessation only occurred in about one-fourth of the severest sufferers, it occurred in about two-thirds of those who did not suffer at all. On studying the thirty-nine cases, I was struck by their presenting a singular absence of all menstrual disorders. In many the menstrual flow began suddenly, continued regularly through life without disturbance, and ended suddenly.

The current belief in the dangers of sudden cessation is grounded on some isolated cases in which it proved dangerous, owing to the action of other causes : thus, in two plethoric subjects, hemiplegia was induced ; but alone it will not cause diseases of cessation, because the system is equally suddenly relieved by some critical discharge, and most frequently by well-sustained perspirations.

VII. Disuse or abuse of the reproductive organs.— Gardanne and other writers countenance the popular prejudice that the unmarried are most liable to flooding, to cancer, to ovarian and uterine tumours at the change of life, but there is no truth in the assertion. The single pass through this period without more trouble than the married, and, so far as I can ascertain, suffer less than other women. Gardanne likewise asserts that prostitutes suffer much at this epoch, an assertion utterly groundless, for Mr. Acton tells us they do not carry on their infamous pursuit for more than three years, on an average; so that they have seceded from the ranks of infamy, long before the change of life. I have occasionally traced an aggravation of diseases of cessation to sexual intercourse, and to women marrying while the change was progressing, I have repeatedly known it to be followed by flooding, by ovario-uterine irritation or inflammation. In one case a sensible lady, whose family was not tainted by insanity, had a flooding on the wedding night, very violent abdominal pains during the following days, and went out of her mind for many months.

VIII. Social position.—I believe with Dusourd, whose practice lay in an agricultural district in the south of France, that peasant women suffer little at this period. They belong to the *genus inirritabile*, and are therefore little liable to nervous disorders. Their health is generally good when the change of life comes on; flooding is frequent, but effectually critical, and they take little heed of the flushes and perspirations which annoy the spoiled daughters of civilization. The poor of large towns suffer much at this epoch. The necessity for working hard, the anxieties of poverty, the impossibility of escaping the annoyances of children, even for a few hours in the day, increase the sufferings of the poor at the change of life. Certain occupations have a similar effect; thus, Auber says that washerwomen suffer more than others, on account of the vicissitudes of temperature to which they are exposed. The close, damp, and heated rooms in which book-folders, catgut makers, and others are obliged to work, increase the sufferings of many; but by a fortunate compensation, the necessity for working hard prevents and cures the nervous affections which so frequently assail the rich at this period; for luxury is the hot-bed of nervous affections, they grow there in profusion, and run into such strange eccentricities, that pathologists have given up the hope of completing their catalogue. Of what use is leisure to practise all the appliances of hygiene, without the resolution to use them? Many of the poor are not forced to work in atmospheres so injurious as those of the heated ball-rooms frequented by the wealthy.

In going through my numerous cases, to discover why some women suffer so much and others so little at this period, I come to the conclusion that it does not so much depend upon the strength, original or acquired, that the system enjoys, and which is *constitution*; nor on the visible predominance of one set of organs over the other, which is *temperament*. Neither does it depend on the menstrual flow ceasing early or late, nor on women being single or married, rich or poor, but on a peculiar susceptibility of the nervous system—a condition which, though hidden from the microscope, is evident from the manner in which it responds to the reproductive and all other stimuli. Women who suffered much at the change of life had often suffered much at puberty and at menstrual periods ; while these had seldom been attended with distressing symptoms in women who suffered moderately at cessation ; and amongst the thirty-nine cases where there was no suffering, there was a similar immunity at puberty and the menstrual epochs. I, therefore, conclude that the diseases of the change of life, like those of puberty, are principally to be ascribed to the nervous system being unable to tolerate the stimulus imparted to it by the reproductive organs, for when the nervous system is well-tempered, this stimulus acts as an improver instead of a disturber of health. Vital acts, however, are never found cut and squared with mathematical precision ; there are, therefore, exceptions to this rule, and I have had patients who suffered much at cessation, whose previous good health had been uninterrupted. These remarks are a fit introduction to what I have to say respecting the theory of the production of diseases at the change of life ; for of what use is a long enumeration of medical facts unless it lead to a sound theory, explaining their production and preventing their occurrence ? Stones only cumber the ground till the architect puts them in place.

Blood Theory.—The more abundant quantity of blood, or its defective qualities, will account for congestions of organs, for hæmorrhages, and for other flows, and for thirty-two years the womb is a powerful centre of attraction, causing it once a month to be the focus of blood-currents from all parts of the body. A strong mental or pathological perturbation sometimes so divides the blood-currents, that while a portion is directed to the womb, the rest is sent into some weak organ, which becomes congested, if unable to throw the blood off from its surface. The same occurs during the change of life. Part of the blood is still every month directed towards the womb, which is thereby congested, while disease may be induced by the monthly current setting in towards a weak organ. After three or four ounces of blood have been retained for a few months, plethora may be induced, if the nutritive force is unable to use the blood to strengthen the interior framework, or to produce fat. Plethora once induced, the system is oppressed and rebels; and if constitutional strength be equal to the task, it throws off the blood, but if strength be below par, the blood stagnates in congested tissues.

Nerve Theory.—It is, however, unreasonable to attribute to plethora most of the cerebro-spinal and ganglionic affections of the change of life. Thus formerly, when women suffered from headache, giddiness, drowsiness, and dulness of intellect, they used to be bled, cupped, or leeched; but though this nervous condition, which I call pseudo-narcotism, may co-exist with plethora, it is not proportionate to it, and may often be found in women who have become chlorotic at cessation, as well as in those whose amount of blood is well apportioned to the wants of the system. To explain, there-fore, the occurrence of all the diseases of the change of life, it is necessary to take into consideration the morbid condition of the ganglionic nervous system during the first and second parts of this period. The condition of the ganglionic system cannot be materially altered without inducing cerebral dis-turbance, which must be admitted before the diseases of women can be understood. In searching how it is that ovarian dis-turbance induces morbid action, its effects on the female economy should be studied, not only in the menstrual pheno-mena, but also in the continuous action of the ovaries on the system. The results of this influence, when morbidly exerted, are often evident long before the menstrual flow, and long after cessation. When a girl of ten years of age is, more or less, in a constant state of pseudo-narcotism, without any sign of the well-recognised forms of cerebral or gastric disease, the premature morbid influence of the ovaries on the cerebral system is manifest, and Landouzy has rightly sub-

mitted that the ovaries may also produce hysteria, long before
the appearance of the first menstrual flow or the comprehension of sexual ideas.   I have seen girls affected with pseudo-
narcotism eight years previous to first menstruation, and
women to suffer severely from it ten years after cessation.
When, therefore, I do not find it explained at the change of
life, by gastric, hepatic, or well-known cerebral disease, I
consider it indicative of perverted ganglionic influence, the
most frequent result of disturbed ovarian action, not only at
puberty and cessation, but at each menstrual period and
during pregnancy and lactation.   This perversion of ganglionic nervous power at the change of life is the result of
increased ovarian irritability during the subsidence of specific
ovarian functions.   In many women, the excessive ovarian
irritability, made evident by so many symptoms, reacts on the
pelvic ganglia, and thereby on the ganglionic nervous centre.
In many, ovarian life is not extinguished, and the ovarian
and uterine ganglia are not condemned to inaction, without
the abdominal federation of ganglia and plexus, feeling more
or less paralysed by the want of that specific influence which
had ruled the system for thirty-two years.   Then occur irregular manifestations of nervous energy and the ataxic nervous
symptoms, which Sydenham dilates on, when treating of
hysteria.   Then occur all sorts of erratic nervous pains, as
well as the anomalous symptoms referred to the pit of the
stomach, and the cerebral disorders of which I shall fully treat.

I have stated, that out of 383 women in whom the menstrual flow had ceased, 223 suffered much, 121 little, and
thirty-nine not at all ; yet I do not consider that the proportion of thirty-nine to 383, or ten per cent., gives a fair idea of
the manner in which the change of life tells upon women, for
the non-sufferers having nothing to complain of, seldom come
under medical notice.   The most frequent forms of disease at
the change of life may be gathered from the following table ;
but the reader will bear in mind that I have been consulted
in many unusually severe cases, on account of my being
known to have given particular attention to the subject.

TABLE XXIII.

Nature of the Affections of 500 *Women at the change of Life.*

Nature of Disorder.	No. of Cases.
Diseases of ganglionic system	406
,, the cerebro-spinal system . .	1272
,, nerves	487
Diseases of the reproductive organs . .	463
Gastro-intestinal affections	354
Cutaneous affections	705
Various affections	43
Total	3730

Each of these divisions will form the subject of a separate chapter, and if the aggregate of suffering puts on a formidable appearance, the reader must bear in mind that in this table, flushing, perspiration, and other slight symptoms count as well as cancer, and that many women were suffering at the same time in many and various ways.

CHAPTER IV.

THE natural history of the reproductive function has enabled me to state in the preceding chapters what complaints are to be expected at the change of life, and will also point out the principles that should govern their treatment.

If nature bled, in different ways, 208 women out of 500, it evidently shows that this spontaneous effort of a hidden force to relieve the system may deserve imitation. If 326 out of 500 suffered from sinking at the pit of the stomach, from fainting, debility, and chlorosis, it shows that stimulants and strengthening treatment must be as indispensable at the change of life as at puberty. If 75 out of 500 had frequent diarrhœa or constipation at this period, it shows that purgatives may often be of great service. If 285 out of 500 had unusual perspirations or sweating, it is a positive proof of the utility of sudorifics at this epoch. If 134 out of 500 suffered continually from biliousness, jaundice, waterbrash, vomiting, and dyspepsia, it is clear that alkalies will be often useful. If in 277 out of 500 the nervous system was actually steeped in a more or less intense state of stupor, it suggests the great utility of sedatives. If, in many, the organ most prone to disease through life suffered most at this period, does it not show the necessity of discovering this weak organ from the patient's previous history, so as to give it protection? If some form of disease of the reproductive organs frequently prevents the change of life taking place after the healthy pattern that I have described, does this not explain that the cure of any such disease is the first step to the recovery of health at the change of life? In all cases the main indications are:—1. To cure local disease. 2. To restrain abnormal discharges. 3. To correct inordinate or irregular nervous action. 4. To strengthen the patient's constitution.

In exhibiting remedies at this period, it is necessary to guard against a prejudice firmly rooted in the minds of many, that the change of life is synonymous with old age, for the principles of treatment applicable to diseases of old age will not always suit those of the change of life. Then, as at puberty, there may be vital energy, but latent and oppressed, so that bleeding and lowering measures sometimes develop an unexpected amount of strength. I thoroughly believe in the efficacy of the modes of treatment thus suggested by the study of natural phenomena, and many cases recorded in this work will show, how, by following the suggestions of nature, I have been able, in a few days, to relieve patients who had been suffering for years. With the exception of those afflicted with cancer or structural diseases, there are few incurable cases, though many, satisfied with a first instalment of recovered health, will not allow a perfect recovery to be made. They would rather bear their accustomed evils than submit to the tedium of following out a systematic plan; they oppose the stubbornness of prejudice to advice founded on fully proved facts, and then talk of the "deplorable inefficacy of medicine," when, in fact, *they will not take the trouble to be cured.*

BLEEDING.—The *natural history* of menstruation informs me that at its cessation, nature has sought relief by hæmorrhages from different parts of the body in 208 women out of 500. As they are often benefited by the loss of blood, the natural process is evidently worthy of imitation, and it has been extensively copied by illustrious medical men of former times, and by Tissot, Hufeland, and Meissner; nevertheless so little does present practice imitate nature, that only five of my 500 patients had been bled, and ten cupped. Fothergill and Heberden thought that sudden deaths had greatly increased in England since bleeding, at the spring and fall, had

gone out of fashion; and whatever may be the truth of this
assertion, I am convinced that the sufferings of some women
are aggravated from bleeding being so seldom resorted to;
notwithstanding its being commended by the facility of
understanding that the well-timed subtraction of a small
quantity of blood from women accustomed for more than
thirty-two years to bleed periodically, is the best means of
preventing or curing some of the diseases originating in the
cessation of menstruation, until the blood can be safely disposed of. Bleeding went out of fashion, because the profession
was vividly impressed with the recollection of the excesses to
which, at different times, it has been carried, and I shall not
discuss the question in its general bearings, as I have done
so in my "Handbook of Uterine Therapeutics." I will
merely remark, that lately, venesection in well-defined cases
has been advocated by good pathologists, like Dr. Handfield
Jones and Dr. Richardson. To those who still retain a horror
of this remedy, I will only observe that, supposing the constitution of *man* has so changed from what it was, that it be
now damaged by taking from him ten or twelve ounces of
blood, when a prey to acute disease, I can answer for *woman*
often bearing with benefit very large losses of blood; and as
a redundancy of blood, of which nature has not yet found
means of disposing, is a cause of disease at cessation, it
follows that bleeding, so often effected by nature, at this period
of life, should not be neglected by those who pride themselves
on understanding and on imitating her proceedings.

Plethora sometimes exists with apparent weakness, which
may be relieved by bleeding; so that in seeking to determine
the utility of this measure, it is well to be guided by the
state of pulse at the temples and at the heart, as well as at
the radial artery, bearing in mind that, should there be much
emaciation, the temporal and radial arteries would be brought
nearer to the surface, and give a first impression of vigour
where none exists. While the effects of bleeding are admirable in many patients of a plethoric type, they would be
very detrimental to those who are chlorotic and nervous.

Characteristic nervous symptoms of the change of life are headache, giddiness, heaviness, and drowsiness. I have drawn attention to these symptoms as of frequent occurrence at puberty, during pregnancy and lactation ; and I find that whatever may be the period of their supervention, they are often considered indicative of a determination of blood to the head. The same symptoms, with headache and giddiness, have been lately considered as indications for bleeding by Columbat and Auber ; and though in England similar mistakes are prevented because nobody bleeds, yet the fashion will change again ere long, and then will occur the danger of mistaking functional cerebral disease for plethora. The following cases will, to a certain extent, exemplify the discrimination required to treat patients suffering under similar symptoms : —

CASE 10.—In 1844, the rage for bleeding, again developed in France by the passionate advocacy of Broussais, had subsided in Paris ; but it was still possible to test the ill effects of this pernicious system. About that time I was consulted by a lady, aged fifty-one, tall, thin, with a pallid complexion, dark hair and eyes. She had first menstruated at fifteen, the function had never been interrupted except by three pregnancies ; it subsided gradually, and cessation occurred at the age of forty-eight. For some months she felt no inconvenience, but afterwards she was much troubled by headache, giddiness, flushes, and perspirations. An eminent French physician, ordered her to be bled to ten ounces ; a slight improvement followed, but the same symptoms soon returned, which were again interpreted as signs of plethora, and other ten ounces were taken from the arm. This second bleeding made the patient worse ; and when I saw her, the marked ill effects of the treatment were very evident, so I gave anodynes, mild purgatives, wine, and a more strengthening diet. The patient rapidly improved, and in subsequent relapses derived benefit from the same kind of treatment.

CASE 11.—In 1850, I was consulted by a lady, aged fifty-three, of middle stature, sanguine complexion, brown hair, and hazel eyes. She menstruated abundantly for the first time in her thirteenth year, and she had since been regular, the discharge being usually abundant. While the function was ceasing, she was twice seized with flooding, and was much better for it. Menstruation ceased at fifty-one, and was soon followed by diarrhœa, which came on at irregular intervals, but did not interfere with appetite and strength. When that supplementary discharge subsided, heaviness of the head, with giddiness, came on, together with flushes of heat and drenching perspirations. For these distressing symptoms she had consulted several medical men, and had taken quinine, acetate of lead, and gallic acid, but without benefit. I ordered her to be bled to twelve ounces, and the vertigo, flushes, and perspirations abated considerably. The bowels were kept open by Seidlitz powders; several glasses of effervescing lemonade were taken in the course of the day, and a tepid bath, for an hour, every week. Meat once a day, no beer nor porter, one glass of sherry at dinner, and increased exercise in the open air. In a month all the painful symptoms had disappeared, and the patient remained well for several subsequent months, when, without any apparent cause, the same symptoms broke out again. I ordered six ounces of blood to be withdrawn, and prescribed the former treatment, with similar good effect.

The symptoms experienced by both these patients were similar, but their constitutions were very different. The first shows that, when nervous people are bled to excess, there arises a state which often closely resembles the threatenings of disease in the vital organs, relieved in other temperaments by bleeding. In the last case the patient was of a strong constitution, accustomed to lose considerable quantities of blood, and relieved by the occurrence. The vigour of the circulation was well proved, by the strong impulse of both heart and pulse, instead of the flaccid condition of both in the first patient. The one was relieved by sedatives and a strengthening diet, the other principally by bleeding; and I have seen bleeding remove these symptoms when it was not indicated by a strong constitution, but by the previously-contracted habit of losing a considerable quantity of blood ; as in women of a slender make and slight delicate appearance, and in those whose nervous susceptibility is great, in whom we must admit a tendency to hæmorrhage.

At first, sufficient blood should be taken away from a plethoric woman to make a decided impression on the system, for nature frequently adopts this plan, insomuch that 138 women out of 500 were flooded at the change of life ; but when the indication to bleed recurs, it is better to bleed in progressively smaller quantities, and at progressively longer intervals. Tissot mentions a case, in which he thought right to bleed for three years, after which the patient recovered her health. Hufeland used to bleed three times in the first year after cessation, twice in the second year, and once in the third. I sometimes follow a somewhat similar plan, which is a daguerreotype of a natural process, for in 171 women out of 500, the menstrual flow ceased naturally, that is, by a gradually smaller amount of discharge, occurring irregularly every two, three, four, five, or six months. In 53 cases out of my 500, there was a marked return of monthly phenomena after cessation. In such cases it would be judicious to bleed before the accustomed monthly occurrence.

The effects of the bleeding should be aided by judicious regimen ; for, doubtless, the necessity for bleeding, even plethoric women, would be considerably diminished, if it were not so difficult to persuade them to break through accustomed habits, and if they would consent, for a time, to diminish the quantity of their food, and refrain from what may be otherwise prejudicial in their mode of life.

LOCAL DEPLETION.—When prejudice interferes with bleeding, a few leeches behind each ear, or cupping at the nape of the neck or between the shoulders, may be resorted to. If piles have formerly had a tendency to bleed, leeches may be applied to the anus ; they will also be sometimes useful if there be habitual congestion or inflammation of the womb, though I was surprised to find Dr. Ashwell advocate bleeding the womb by leeches as the best mode of depletion at the change of life, for leeches are generally applied to the womb so as to determine blood to that organ; and to apply them at this period, except under peculiar circumstances, would be likely to prolong what nature wants to curtail. To check the determination of blood to the womb is clearly indicated at the change of life; and even in cases of uterine inflammation I seldom apply leeches, for I find that repeated small general bleedings are more effectual in checking that monthly turgescence of the womb which may take place long after cessation.

PURGATIVES. — Diarrhœa relieved fifty-two out of 500 women at the change of life. These bear witness to the utility of purgatives, as well as the thirty-seven who were subject to dyspepsia, the twenty-three who were troubled with obstinate constipation, the fifty-five in whom biliousness was a prominent ailment, and the six who had jaundice once or more frequently. The utility of purgatives at cessation is further shown by the very intimate sympathy which I have proved* to exist between the generative and intestinal canals, and which is indicated by the relaxation of bowels which generally attends the menstrual flow. Purgatives constitute one of the most convenient local depletions by which the system can be habitually relieved; their employment, therefore, at the ménopause, fully deserves the confidence which the profession has long accorded to the practice, and the sanction which Fothergill has given to it; indeed the utility of purgatives has become so much a matter of popular, as well as of medical belief, that both patient and medical attendant too often confide in these alone, to the neglect of other important means. I might give numerous cases in proof of the utility of diarrhœa, but I will merely state the following.

CASE 12.—Catherine M., aged fifty-three, tall, thin, and

---

* "On Uterine and Ovarian Inflammation, and on Diseases of Menstruation." 3rd edition.

pale, menstruated very abundantly at fifteen years of age, was regular from the first, and continued so for three or four days every three or four weeks, with so little suffering that " she never felt them come nor go." She married at thirty-three, miscarried three times, and bore five children, the last at forty-seven; and menstruation, which had been irregular a year previous to conception, never returned after that event. The patient was generally relaxed during the menstrual epochs. During her last pregnancy, and after her confinement, she frequently had three or four stools a day, without pain or loss of appetite, and since then, diarrhœa came on every three or four weeks, with flushes and drenching perspirations. For the last twelve months she was relieved six or seven times a day, until lately, when this had only occurred once in two days, and she has suffered much from heat, flatus, nausea, oppression at the pit of the stomach, and want of appetite; her tongue being always clean and healthy. When the action of the bowels became freer, the patient got well.

When diarrhœa occurs at the change of life, it is generally at irregular intervals; it may, however, appear with the regularity of the menstrual function, as in the instance of a lady, forty-five years of age, in whom, at the accustomed time, diarrhœa came on when the menstrual flow was due. The diarrhœa lasted three days, and gave relief, although it was not followed by the menstrual flow, which never returned, and was thus replaced by the recurrence of diarrhœa every month for a year, and with great benefit to the patient. I have known the cessation of menstruation to have been followed for five years by an habitual looseness of bowels, occurring two or three times a day, generally without colics. The patient enjoyed good health during that time, and is now a stout and tolerably healthy woman. Dr. Day notices the salutary effects of diarrhœa, consisting of watery evacuations, taking place without apparent cause every three or four months after the cessation of menstruation; and he mentions the case of a lady, eighty-seven years of age, in

H

whom this had happened with great advantage for the last
thirty years. Instead of giving drastics it is more prudent
to prescribe the frequent use of the milder opening medicines,
which may diminish by degrees abdominal plethora, the
more so as it may be many months before the constitution
can settle down. The cooling saline purgatives serve this
purpose, such as the soluble cream of tartar, cream of tartar
lemonade, citrate of magnesia, Seidlitz powders, artificial
Cheltenham salts, or Epsom salts in small quantities, Pullna
and Friedrichshall water, a wineglassful being taken on
rising from bed. I occasionally prescribe the soap and
Barbadoes aloes pill of the British Pharmacopœia, ordering
five or ten grains to be taken with the first mouthful of food
at dinner.

Sulphur is generally classed amongst purgative remedies,
because such is its visible action; but it owes its chief value,
in diseases of cessation, to another action, much more difficult to understand, and which has long rendered it so
valuable both in hæmorrhoidal affections, where there is an
undue activity of the intestinal capillaries, and in skin
diseases, morbid phenomena that may be expected at the
change of life. Sometimes I administer the flower of sulphur
alone, or else to each ounce of it I add four drachms of
calcined magnesia; at others, I prescribe equal parts of
borax and sulphur, one to two scruples of these powders to
be taken at night in a little milk, which generally acts
mildly : and such combinations are very valuable when a
continued action is required. Thus taken in small quantities,
and now and then left off, I have not heard sulphur objected
to by patients on account of its determining any peculiar
smell. A remedy called the Chelsea Pensioner, of which Dr.
Paris has given the following formula, agrees well with some.
Of guaiacum resin, one drachm ; of powdered rhubarb, two
drachms ; of cream of tartar and of flower of sulphur, an
ounce of each ; one nutmeg finely powdered, and the whole
made into an electuary with one pound of clarified honey : a
large spoonful to be taken at night.

ALKALIES AND DIURETICS.—In healthy women, the monthly ovarian crisis generally produces *critical* deposits in the urine, for most women remark their water to be "muddy" a day or two before or during the menstrual flow. During the dodging time and after cessation, this turbid state was noticed in forty-nine out of my 500 cases, to last for weeks, to disappear and to return again. This is sufficient to indicate the utility of alkaline preparations; their utility will be likewise apparent in the thirty-seven cases of dyspepsia, the thirty-one of vomiting, the five of waterbrash, the fifty-five of biliousness, and the six of jaundice. Dr. Parkes has shown that the action of liquor potassæ on healthy subjects varies, according to whether it be taken before or after meals. If taken after meals, the liquor potassæ acts as an antacid. It combines with the hydrochloric or lactic acid, and passes into the circulation without increasing the water, the solids, or the sulphuric acid of the urine, improving digestion and also the state of the blood. If, on the contrary, liquor potassæ be taken before meals, it has the power of reducing *embonpoint*. From thirty to ninety minutes after the liquor potassæ has entered the circulation, there is an increased flow of slightly acid urine, which contains the whole of the potash and organic matter, and a relatively large proportion of sulphuric acid. In other words, an albuminous compound either in the blood itself or in the textures becomes oxidized; its sulphur, under the form of sulphuric acid, unites with the potash, and possibly with the changed protein compound, and is eliminated by the kidneys. The amount of albumen or fibrin thus destroyed by a few doses of liquor potassæ is doubtless small, but as the remedy can be taken for a considerable time, and its oxidizing effects can be assisted by exercise and by copious draughts of water, there is a possibility of removing superfluous matter from a patient without risk. At all events some women who became very stout after cessation, and who derived little benefit from measures advised to relieve the sweats and nervous symptoms of this period, rapidly improved, when, by large doses of liquor potassæ, they had been disencumbered of a superabundant amount of fat.

Dr. Shearman has drawn attention to the fact of the urine being saccharine in some of the forms of ganglionic disease, to be treated of hereafter, and which he considers dependent on neuralgia of the vagus nerve, and he places as much confidence as I do in alkalies, for he gives ʒj doses of liquor potassæ to determine an alkaline condition of the blood, to reduce the sugar in the urine, and bring it back to its normal state of acidity. In many diseases of cessation, alkaline medicines are required as antacids and blood improvers; and I frequently order a tablespoonful of a six oz. mixture, containing, amongst other ingredients, two drachms of liquor potassæ, to be taken half an hour after meals. Unless there be much flatulence, or the weather be cold, an effervescing draught is a convenient way of giving alkalies; and borax, acetate of potash, and nitrate of potash may be added when the object is to act on the kidneys.

Mineral Acids.—The frequent occurrence, at the change of life, of dyspepsia, biliary derangement, and debility, makes it often desirable to give mineral acids, and I prescribe thirty drops of the diluted nitro-muriatic acid in distilled water or other vehicle; and sulphuric acid is useful whenever the functions of the skin are disturbed.

SEDATIVES.—The 227 women out of 500 who suffered from pseudo-narcotism, the 226 who were troubled with dorsal pains, the 205 distressed by hypogastric pains, the 208 subject to headaches, the 146 presenting the minor forms of hysteria, the forty-nine distressed by epigastric pains, the sixteen in whom insanity occurred, and many others who suffered from less frequent nervous affections, are all witnesses to the efficacy of sedative medicines, in relieving the many forms of nervousness unavoidable at the change of life. The utility to be derived from sedatives in diseases of old age has been so admirably pointed out by one who has thrown considerable light on their difficult study, that I avail myself of Sir H. Holland's remarks, which apply forcibly to diseases of the change of life.

"The manner of employing opium in modern practice might, until very lately, be cited among the many examples of perverse changes of fashion as to particular remedies and methods of treatment. The fear of confining the bowels and checking the secretions, constantly present to the mind of the practitioner, readily imbued the patient with the same alarm ; and thus far prevented the adequate use of a medicine having power of mitigating pain, of relieving spasm, of procuring sleep, of producing perspiration, and occasionally even of aiding the natural action of the bowels, by obviating the disordered actions which interfere with it. I speak of this as having been, because it is certain that opiates are again more largely employed, since the introduction of morphia as a common preparation has furnished new methods of administering the remedy, and revived attention to the principles of its action. Yet even now it may be affirmed that there exists a distrust, both as to the frequency and extent of its use, not warranted by facts, and injurious in various ways to our success in the treatment of disease. This is the more singular, seeing the boldness of our practice in other points, that we have in the sleep produced a sort of limit and safeguard to its effects, and that we possess remedies of easy application for all injurious symptoms that can arise. To the insufficiency, indeed, of the quantities given, may be attributed, in some part, the comparative disregard into which the remedy fell during a certain period. Half a dose might disturb and distress the night which a full dose would have made one of perfect rest, or perplex the aspect of symptoms which a larger quantity would have alleviated or removed.

" Yet medical experience does but follow common observation in recognising the inestimable value of sleep in sickness, of the suspension of pain, and the check to all disordered actions thereby obtained. For pain and sleeplessness, though strictly but symptoms of other ailments, may often, in practice, be viewed as disorders in themselves, the removal of which is essential to the success of our general treatment. How frequently do we see a nervous restlessness come over the patient, the consequence of protracted sickness or other causes, retarding cure by preventing the due effect of remedies, and receiving no relief itself from the means employed for the disease? In such cases, the physician is not to submit himself to names or technicalities. The regular course of treatment must be suspended until the hindrance is removed; and even seeming contradictions to this course may safely be admitted for the attainment of the object. Here opium is the most certain and powerful of the aids we possess; and its use is not to be measured timidly by tables of doses, but by fulfilment of the purpose for which it is given. A repetition of small quantities will often fail, which, concentrated into a single dose, would safely effect all we require."

These remarks are in every way applicable to diseases of the change of life; for sedatives, by assuaging the acuteness of pain, lull excited action to a slower rate of progress, and to a more subdued tone, for the blood-vessels serve under the vaso-motor nerves; and sedatives, by restraining the heart's action, diminish the momentum of the blood. Moreover the principal source of cerebral disturbance at the change of life is, that irregular stimulus of the reproductive organs, which are no longer relieved by the menstrual flow; indeed some of my patients have been driven to the verge of insanity by ovario-uterine irritation; and in proportion as this was assuaged by the systematic exhibition of sedatives, cerebral disturbance abated. Those most benefited by sedatives are the nervous and chlorotic, in whom there is often action without power—action requiring to be restrained until the system has gathered strength; and as the tolerance for a remedy is generally in direct proportion to the possibility of its being required, it will be found that the thin, weak, and nervous are most sensitive to the action of sedatives, and can be brought to tolerate the largest doses. Those of the plethoric type bear the solanaceous sedatives better than opium, unless this be given with ipecacuanha and purgatives; and calomel or blue pill should be combined with the sedatives when the liver is out of order. It stands to reason that at the ménopause sedatives may be administered by the mouth, by means of hypodermic injections, and by topical application to the disturbing seat of morbid action. Opium and belladonna plasters, and, better still, those made by incorporating morphia and atropia with glycerine ointment, according to the method explained in my "Handbook of Uterine Therapeutics," are singularly useful to quell the epigastric pains that women suffer from so much, and which play so important a part in the nervous affections of women. The precept of applying the sedative as near as possible to the suffering nerves explains the speedy and beneficial results that often follow the exhibition of suppositories or medicated injections, by the vagina or the rectum, when the reproductive organs are diseased. Even in cases where there are no abdominal or lumbar pains, I sometimes order sedatives by the rectum, with the greatest benefit. In some cases of nervous irritability, where there is a great tendency to headache, delirium, pseudo-narcotism, or any state approximating to insanity, they are most useful ; for at the change of life all those distressing conditions of the mind do not depend upon idiopathic cerebral disease, but on the perverse reaction, on the brain, of some anomalous ovarian influence transmitted through the ganglionic nerves, and by assuaging the pain, irritation, and undue influence of the reproductive organs, sedatives, thus given, allay the disturbed action of the ganglionic nerves, and thereby cure cerebral affections.

With regard to the special indications for sedatives, to avoid repetitions, I shall refer the reader to the following chapters, in which the treatment of each disease is given, and to my " Handbook of Uterine Therapeutics and of Diseases of Women." I will not, however, dismiss the subject of sedatives without a general survey of their relative value.

To relieve the slighter forms of cerebral and ganglionic disturbance there is nothing like henbane, for it is almost always more or less effectual, and produces no untoward

effects, even if continued for a long time. I have less experience of conium, but believe it to be almost as useful.
Indian hemp, when it agrees, and it often does, has the great
advantage of not producing constipation or heating the
system. Belladonna and atropia act admirably in suppositories or plasters, but I rarely give them by the mouth.
Concerning the utility of opium and its numerous components,
I cannot speak too highly, whether as a topical agent or as
the means of influencing the whole system. I first try
Dover's powder, and then morphia, beginning of course with
small doses, and much as I deprecate its continued use in
large doses, I have not hesitated to sanction the daily taking
of two and three grains of acetate of morphia, for several
years, when I found that nothing else could quell sickness,
diminish ovario-uterine pain, cure hysteria, and render
bearable an existence otherwise intolerable. Such cases are
very rare, but they must be allowed the benefit of the
habitual use of exceptionally large doses of morphia.

Hydrocyanic acid in large doses is only suitable when
strength is above par, but in small doses I occasionally find
it to be better borne than other sedatives by nervous women.
Although Dr. Physick exaggerates when he says that " camphor was made for women, with whom it always agrees,
while it always disagrees with men," it is a very valuable
remedy. The subtle fumes seem to spread like an aura over
the nervous system, stimulating it to increased action, causing
the capillaries to eliminate with the perspiration, whatever
oppressed the nerves, whether it be a liquid, gaseous, or an
electroid fluid with which they were overcharged. The effect
of this is a subsidence of pain, an increase of strength, and
sometimes a sensation of lightness. Camphor seems to
correct the toxic influence which the reproductive system has
on the brain of some women, and its anaphrodisiac properties
have been often shown in priapism and nymphomania. It
abates the sexual sting by acting on the cerebro-spinal nerves
of the external organs of generation, not on the testicle or
ovary. The testimony of Raspail on this point is of great

value, and he says, that habitual large doses did not prevent conception nor induce impotence. Camphor may therefore be useful in appeasing the excitement of the organs of generation at the change of life, if it be given in sufficiently large doses. Lupulin has likewise a strong anaphrodisiac action. Castor, ambergris, musk, and sumbul are nervine stimulants that are often useful, castor being the one I most frequently give, and I find it is the principal ingredient of Crollius's Uterine Elixir. Chloral will, I believe, be invaluable in nervous diseases of women; its uniform action, without causing constipation, or nausea, or depression, recommends its adoption. In some cases of perverted cerebral innervation the inhalation of chloroform, or of a mixture of chloroform and ether, may be usefully carried to the verge of unconsciousness. I cannot dismiss this subject without repeating a hope expressed in successive editions, that sedatives may be more systematically used at the change of life than is now the case, so as to enable women to withstand the over-exciting effects of the present civilization on the nervous system, and to deaden the reproductive stimulus, which only lingers on to disturb health.

SUDORIFICS.—In some countries the polite question on meeting a friend is not, " How do you do?" but " How do you perspire?" This might apply to women at the change of life, for the 201 out of 500 who had continual perspirations, and the eighty-four who were for a long time repeatedly sweated, show the utility of sudorifics. Indeed, for one woman who is for a time relieved by a critical exhalation of blood from the womb, twenty, or more, derive permanent benefit from long-continued critical cutaneous exhalation, so I wonder that the utility of perspiration, and the advantages to be derived from sudorifics, have not been more insisted on, especially as their action is exerted on an immense surface, and followed by the exhalation of a large quantity of water, salts, and animal matter. That the perspirations of this period have not passed quite unnoticed is evident from Siebold mentioning a case of sudden cessation, followed by abundant perspiration. Tissot also alludes to the fact, that when intermittents were frequent, he sometimes saw menstruation finish by an intermittent fever, which, by greatly augmenting the perspirations, relieved the patient of the other infirmities which so often attend this critical change. For proof of the utility of perspirations in many affections at the change of life, I refer to many cases in this work. Siebold's case reminds me of one in which the menstrual flow left suddenly at forty; since then she never suffered, but had constant and gentle perspirations up to the fifty-fifth year of her age. Eliza S., aged fifty-two, lost the menstrual flow suddenly at forty-two, after venesection, during a menstrual period; since then, she has had constant gentle perspiration, but no suffering. On the other hand, whenever women suffer more than usual from distressing sensations referred to the pit of the stomach, and from a considerable amount of cerebral disturbance, it will frequently be ascertained that the skin is habitually dry. There are "dry flushes," in which the skin is not relieved by perspiration. In such cases, profuse perspiration, whether an effort of nature, or the effect of medicine, brings speedy relief. To induce perspiration, I sometimes begin with an emetic; more frequently a tablespoonful of acetate of ammonia given repeatedly, the patient being in bed, is useful: when there is no tendency to cerebral congestion, I have given vapour baths, though I generally prefer the more permanent gentle perspiration brought on by a cooling regimen, by warm baths and sulphur, of which I give one or two scruples once or twice a day.

BATHS.—The 287 women who suffered much from repeated flushes, the eighty-four who were habitually sweated, and the forty-two who presented various cutaneous affections, bear witness to the utility of baths. A warm bath is a very simple thing, but it acts in many ways. 1. It removes from the skin the saline deposits and other secretions left there by continued perspiration. 2. The veins of the skin absorb a certain amount of water, which allays cutaneous irritation and dilutes the blood. 3. The warm bath is a gigantic poultice, applying its warmth to all the peripheric expansions of the nervous system. In some hidden way the warm bath is a positive corrective of nervous irritability and a sedative to the nervous system; and when it is considered that this gigantic poultice is perfectly manageable—that the temperature may be increased or lowered according to the patient's feelings and the practitioner's discrimination; that while the patient is in warm or tepid water, cold applications may be made to the head, and a stream of cold water directed to the abdomen through a vulcanized india-rubber tube, so as to quell local congestion, it is not surprising that the greatest men have, at various times, vaunted the utility to be derived from baths in nervous affections, and that, lately, they have been revived in France for the treatment of mental diseases. As a general rule, I tell my patients to take a warm bath every week, and to stop in the water for an hour. The prejudice against baths, except as a means of cleanliness, is so great in England that I am afraid of asking patients to do more, but when I wish to obtain the full sedative effects of baths, I direct highly nervous patients to take them, heated to about 93°, and to remain in one, two, or three hours, having warm water added at times so as to maintain a grateful temperature. A greater sedative effect will be obtained by letting the water gradually cool down to 90°, or even lower, and the amount of water absorbed will be to a certain extent proportionate to the low temperature of the bath.

MINERAL WATERS.—Cold sea-bathing is out of the question at the change of life. I have not advised warm sea-baths, but several patients have tried them without deriving benefit, while others have thought themselves worse for their use. Gardanne states, that the effects of mineral waters are rather disadvantageous than otherwise; but he does not say how or where they are taken, and it stands to reason that if there be truth in what I have just stated respecting the value of sulphur, alkalies, purgatives, and steel, in the treatment of diseases at the change of life, these remedies must be likewise useful when combined in mineral waters to be taken internally or used as baths, the more so as these mineral agents exist in a state of combination with other substances so as to be inimitable by our chemistry. Besides this important consideration, the total freedom from domestic cares, the complete change of air, food, habits, and scenery must enhance the value of mineral waters. Sulphurous waters may be good, not only in cutaneous affections, but in many cases of congestion of the internal organs, and in anomalous ganglionic neuralgia, alkaline waters like Vichy or Ems, cannot be overrated in cases of obstinate biliousness, dyspepsia, vomiting, and gout. The saline waters of Homburg and Kissingen serve in cases of dyspepsia and obstinate constipation. Chalybeate waters, like Schwalbach and Tunbridge Wells, are useful in chronic debility, chlorosis, and in the after part of the treatment of many of the gastro-intestinal affections. Hot sulphurous baths, like those of Aix in Savoy or Harrogate, may be made serviceable in the cure of rheumatic affections, lumbo-abdominal neuralgia, local paralysis, paraplegia, and in most neuralgic affections. To whichever of these watering-places the patient may be told to proceed, she should never begin a course without consulting one of the resident medical men, who will be best able, by local knowledge, to advise her respecting the kind of mineral water to be taken, as well as the best mode of its administration in her case.

Issues.—I should not mention these if Dr. F. Churchill had not expressed his agreement with Fothergill, as regards the utility of applying issues and blisters to those who, in early life, have been relieved from cutaneous or other disorders by the establishment of the menstrual flow. He says, " I have repeatedly tried caustic issues, or perpetual blisters, and with the greatest advantage. They certainly aid the action of the remedies already mentioned, and, I think, prevent the recurrence of those irregular congestions which Dr. Fothergill has described." Gardanne and B. de Boismont likewise speak confidently on the utility of issues in preventing disease at the change of life, but I agree with Dr. Ashwell that they are seldom necessary, except when cutaneous eruptions are troublesome, and have been so in youth.

In perusing this work the reader will bear in mind that the 500 women of whom I have traced the varied modes of suffering at the change of life only constitute a minority when compared to those who pass through the change with little or no suffering. Well regulated hygienic habits are all they require to assist the silent operation of natural laws, and I shall therefore treat of hygiene in the next chapter.

CHAPTER V.

As at puberty, from the total ignorance in which it is still thought right to leave young women, so at the change of life, women often suffer, from ignorance of what is to occur, or from exaggerated notions of the perils that await them. It would be well if they were made to understand that, if in tolerable health, provided they will conform to judicious rules of hygiene, they have only blessings to expect from this critical period. The change of life may be dangerous for those who are always ailing, for habitual sufferers at the menstrual periods, and for those affected with uterine diseases; and according as the sufferings of women were protracted, previous to the healthy establishment of the periodical flow, so may they expect its cessation to be attended by a corresponding period and intensity of suffering. Women should know, that unless they be pregnant or nursing, great irregularities in the monthly appearance, or its prolonged absence, coinciding with sensations of sinking at the pit of the stomach, with flushes and perspirations—even though their age may only be between thirty and forty—may, in general, be considered as warnings of cessation, particularly if they are accompanied by a corresponding amount of pseudo-narcotism. This knowledge would prevent cessation being considered in the light of temporary suppression, and forcing medicines and purgatives being taken without the sanction of medical advice. If, on the first indication of the change of life, women who are in fair health sought advice, carefully followed a regimen, and pursued a line of life in harmony with the physiological processes on which this change depends, I believe that almost all disease would be prevented; but as it is the end of a natural function, it is thought right to leave it to nature; no additional precautions are taken, and advice is sought for when the mischief is done. It is, then, well worth while to place on record the rules of hygiene best calculated to prevent and to cure the diseases of the change of life, referring these rules to the great functions of the human economy.

Hygiene of the reproductive organs.—I have shown that there is something more or less anomalous and morbid in the reproductive organs and their action on the system during the first half of the change of life, and that in the latter part of this period these organs have a tendency to become atrophied. Can there be a clearer indication that until after the ménopause, their hitherto appropriate stimulus interferes with a natural process? Hence it is unreasonable to marry during this unsettled period. Experience, moreover, teaches me, that even in those who have been long married, connexion at the change of life is a cause of uterine disorders, and that these have frequently occurred in women marrying during this epoch. I admit with some physiologists, that, as a flickering flame gives a final blaze, so in some women, sexual desire is strongest when the reproductive power is about to be extinguished : this, however, is not the rule, for I have been repeatedly made aware that a distaste for connexion was the first sign of an approaching change. I therefore believe a marked increase of sexual impulse at the change of life to be generally an anomalous if not a morbid impulse, depending upon either neuralgic or inflammatory affections of the genital organs, thus corroborating B. de Boismont's assertion, that " whenever sexual impulse is first felt at the change of life, some morbid ovario-uterine condition will be found to explain it in nineteen out of twenty cases." At all events, I deem it imprudent for women to marry at this epoch without having obtained the sanction of a medical adviser. If this had been done in cases that have come under my observation, flooding would not have followed marriage, slight uterine disease would not have been considerably aggravated, the march of undetected cancer would not have been hastened, and others would not have become insane, as in the following case :—

CASE 13.—Mrs. B. was fifty-one when she consulted me; she had been all her life an intelligent, active, and determined woman, eccentric but not nervous; and when about fifty, and during the dodging time, she married. In the wedding night she had severe abdominal pain, and flooded to a great extent during the following days, and then her mind became affected, with occasional melancholy fits and suicidal tendencies. It was remarkable how her mind was infected with doubt; she could not make up her mind to anything, and was always doubting about right and wrong. When I first saw her she had not slept for a fortnight, and had sensations of burning in the breasts and the lower part of the abdomen, menstruation being irregular, scanty, and dark-coloured. I gave morphia internally, ordered a succession of belladonna plasters to the pit of the stomach, effervescing draughts and warm baths. She got very much better in a fortnight, and was well in about six months, menstruation having ceased. This occurred in 1847, and since then there has been no return, notwithstanding the long illnesses and the deaths of her husband and mother.

The following case also shows how women sometimes seek to stay the inexorable hand of time by protracting the regular appearance of the sign of womanhood; but should they succeed in bringing on the flow, it arises from a diseased state of the womb, and can give no hope of progeny.

CASE 14.—Some years ago I was consulted by a lady, aged forty-eight, who, when about twenty-five, formed a strong attachment, but family circumstances prevented a union taking place, till fortune smiled upon the parties, and the lady married at forty-five. The discharge had proceeded regularly as to time, quantity, and symptoms, up to the period of marriage, but subsequently it never reappeared. As this sudden cessation coincided with gastric symptoms, with a distension of the abdomen, and, above all, with a great anxiety for children, the patient was considered pregnant, and carefully watched for many months. When the illusion was destroyed, the lady became disconsolate; and, punctilious in her notions respecting honour, she brooded over the possibility of her husband supposing that the courses had stopped previous to her marriage. After a minute investigation, I intimated my conviction that health was perfect, and that the monthly flow had ceased. About a year afterwards, I was asked to see her again, and learned that, having consulted some one else, she had taken steel, purgatives, and large doses of savine, until she had a terrific flooding. She subsequently had a continued sanguinoid discharge for several months, and other symptoms of inflammation of the body of the womb, which was most likely caused by injudicious treatment, as well as by marrying when the change was impending.

The preceding remarks are not uncalled for, since it appears from the Registrar-General's report, that in 1851, 982 spinsters and 2375 widows married from the forty-fifth to the fifty-fifth year. If 3357 English women, every year, do what seems to me injudicious, it is that human beings are not governed by physiological considerations, and that women marry to better themselves, or to show that they could have married before, if they had liked; while others, like besieged fortresses, surrender at last for the sake of quiet and peace. When menstruation is well over, there is no reason why women should not marry, although marriage will render them more liable than they would otherwise have been to vaginitis and the milder forms of cervical inflammation.

External irritation is susceptible of being greatly alleviated or cured by lotions of water, or of linseed tea, tepid or cold, ten to twenty grains of acetate of lead, in a pint of spring water may also be used three times a day. Vaginal in-

I

jections and enemata, with tepid or cold water, are beneficial
in allaying irritation, and the following puff powder may be
freely used after washing :—

Camphor, carefully powdered	℥vi
Acetate of lead	℈i
Starch	℥ij
Essential oil of bitter almonds	ℳx

The tepid bath at 92° to 95°, prolonged for an hour once or
twice a week, will be very useful. Repose will permit these
measures to have their full effect, so it would be well to
recline on the sofa for several hours in the course of the day.

Hygiene of the Digestive Functions.—Having shown that
the debility often experienced at the change of life depends
upon a shock felt by the nervous system, or upon the frame
being oppressed by an overplus of blood of which there is no
longer a monthly drain, it is rational that from the very first
appearance of the irregularities which characterize this epoch,
women should curtail, rather than augment, the amount of
food and stimulants to which they have been accustomed.
When in the family-way or nursing, or so long as the menstrual flow remains regular and abundant, many women can,
without inconvenience, take meat three times a day, and
beer and wine at both luncheon and dinner ; but when the
surplus blood, produced by high feeding, can neither be well
employed nor regularly ejected, it increases all the sufferings
of the change of life, and either prevents or diminishes the
efficacy of remedial measures. I should be ashamed of in-
sisting on anything so self-evident, if I were not often consulted by plethoric patients, to whom generous diet had been
recommended to relieve the nervous symptoms of which they
complained. Some had disregarded the advice because they
found high living increased their suffering and their sleeplessness ; whereas when living rather low they slept better and
suffered less. To women of this type, do these observations
apply, and their breakfast should consist of toast or bread-
and-butter, with tea, coffee, or cocoa ; they should take but

one dinner in the day, at whatever hour may suit, but, should it be late, a slight luncheon of bread-and-butter, cake, or biscuits, with a glass of lemonade, ginger-beer, or soda-water, may be taken. Their dinner should be a plain meal; fish, and white meats, such as fowl or veal, in preference to beef or mutton; more of the crust than the crumb of the bread, and jellies and ice in preference to puddings. If, on the contrary, all sorts of farinaceous food, pastry, and cakes, be indulged in, the natural consequence will be a desire to prolong sleep, a distaste for exercise, congestions, bleedings, inflammations, and apoplexy, or they will become distressingly stout. It would be better for both stout and thin to make their principal meal at the fashionable luncheon-hour of two o'clock, in order that their last meal may be light, which can well be managed, even if the usages of society should require a lady's presence at an eight o'clock dinner-table.

Two or three cups of tea, with dry toast or bread-and-butter, may be taken if the dinner-hour be early, and a cup of sago, arrowroot, or something equally light, for supper; but those who dine late should avoid suppers. Ripe fruits and vegetables may be indulged in, to any amount not interfering with digestion. While those who have a tendency to become fat require different food, sleep, and exercise to those who continue thin at cessation, both should beware of giving way to those sensations of internal sinking and exhaustion to which all are equally subject, and not to seek *too often* a comfort for languor, weakness, or nervousness, in wine, cordials, and spirits, by which a temporary support only can be obtained at the expense of an increase in the faintness, flushes, perspirations, and nervous symptoms.

The system requires soothing by medicines and regimen, rather than stimulating by spirits; and I have known instances of women who, by a misinterpretation of their sufferings, have gradually so increased their usual consumption of wine or brandy, as to induce habits of intoxication. One or two glasses of sherry, claret, or champagne, is an average allowance, either alone or combined with seltzer-water; but some have been obliged to discontinue even this small quantity, on account of its affecting the head. Effervescing draughts, such as effervescing lemonade, ginger-beer, &c., are well indicated; coffee should be made weak or avoided, as it is decidedly more heating than tea, and tea should be used in moderation. Patients have prolonged their sufferings by ringing at every hour of the day for tea; for ten or twelve cups of strong tea in the twenty-four hours tell injuriously on the nervous system. Some women of the plethoric type would do well to abstain, for a time, not only from wine and coffee, but also from pepper and spices. Owing to an increase of fat in the omentum and lining of the abdominal walls, as well as to frequent flatulence, the abdomen often becomes very protuberant or pendulous, in which case a light, elastic abdominal belt gives great comfort.

When women are thin, weak, and semi-chlorotic they require more food and stimulants; but although they bear them without inconvenience, or with positive benefit, they should seek advice, for such a state implies the want of medicine.

Hygiene of the Cutaneous Functions.—With a view of preserving health the bath should be used at least every week by those who are at the change of life, and more frequently by those who have profuse perspirations. The water should be sufficiently warm to impart a grateful sensation, from 92° to 94°, the patient remaining in it half an hour or an hour; and if there be any tendency to headache, the head should be sponged with cold water. A cosmetic bath may be made by taking two lbs. of barley, or bean flour, or meal, eight lbs. of bran, and a few handfuls of borage leaves, and boiling these ingredients in a sufficient quantity of water. This both cleans and softens the skin. Long-continued gentle friction of any painful part by a warm affectionate hand is often of great use. While watching in Eastern countries the long-continued and gentle rubbing of the limbs and soles of the feet of the rich by their attendants, I could not help thinking that the benefit might be less due to the mechanical friction than to some kind of influence on the nervous expansions with which the skin is so abundantly supplied. Rubbing until the skin glows, or the actual grooming of the human body, may be very useful to improve the health of scrofulous children, or chlorotic girls, but it is not adapted for women at this period of life. At cessation the vicarious functions of the skin are so important as to require its being covered as much as possible; and when there is a great tendency to perspiration, flannel should be worn immediately over the chemise, and over the nightdress in winter, when the patient perspires profusely in bed, otherwise she may be chilled by any change of posture.

Hygiene of the Muscular System.—Exercise is extremely useful at the change of life, for three reasons. 1st. It relieves the congestion of the internal organs, transferring the blood from them to the limbs, and if the late-hour exercises of civilized life are dangerous to all, particularly to the sanguine, —gentle, regular, and long-continued exercise in the cool of the day is very beneficial. 2nd. It has a depletive effect, causing the skin to perspire more, and the kidneys to excrete more urea. Dr. Leyman found that, by substituting violent for moderate exercise, the relative quantity of urea in the urine varied from $32\frac{1}{2}$ to $45\frac{1}{2}$; and Mr. Simon found that after two hours' violent exercise, the quantity of urea in the urine passed afterwards, was double that contained in the morning urine. This may be one of the reasons why some plethoric patients derive benefit from exercise carried to a certain degree of fatigue, as recommended by Auber. 3rd. Exercise acts by exhausting those redundant energies which, however little understood, when unemployed, produce the fidgets, nervousness, temper, hysteria—a sliding scale which so imperceptibly leads to more serious mental disorders that its successive stages pass unnoticed, until the unforseen climax demands strict inquiry into the patient's previous state.

The observations relative to exercise apply more particu-

larly to walking; driving in an open carriage is likewise
good, but horse exercise should be left off till after cessation;
indeed, its utility in favouring the menstrual flow sufficiently
points to the discontinuance of the practice during the time
preceding and following the change, for then it is likely to
cause flooding, piles, and leucorrhœa. It will be obvious
that the preceding remarks do not apply when the change is
characterized by intense constitutional debility. Walking is
often then an utter impossibility, and should only be gradually
attempted, and riding is out of the question.

Hygiene of the Nervous System.—It cannot be too often
repeated that nervousness, under every shade and type, may
be anticipated at this period, and should be prevented by
suiting the habits of life to the changes progressing. It is
therefore sufficient to mention that the sufferings of this epoch
will be increased by frequenting balls, routs, operas; for, in
addition to the numerous excitements to be there encountered,
hot and impure air must be breathed. The precepts already
given bear indirectly on the abatement of nervousness; but
I can show how susceptible the nervous system is to the
danger of sudden bad news, or to any powerful impression,
by taking my examples from the lower orders, whose nervous
system is less excitable than that of their more favoured
sisters.

M. N. was well and regular up to fifty, when she went
to see a neighbour's child, who was severely burnt, and a
dreadful flooding was the consequence. Hearing suddenly of
the death of her husband brought on a severe flooding in S. L.
In A. B. the menstrual flow suddenly stopped at forty-five,
on her first acquiring the conviction that her husband was
insane; and in several instances, has the news of the sudden
death of a husband caused early and sudden cessation.
Perhaps the most curious effect of the mind on the body, at
this period, was lately published in a French journal. A
highly nervous lady, aged forty-eight, had ceased to menstruate four years, when she attended a sister during a protracted and painful labour. A few hours after it was over,

she was herself taken with similar pains, which produced flooding for several days. Three days after this ceased, the breasts swelled, and a milky fluid came from the nipples. B. de Boismont and others have likewise seen flooding, at cessation produced by strong emotion, or a sudden fit of anger.

Sleep should not be too freely indulged in by those who belong to the plethoric type; thin and nervous women, on the contrary, may be encouraged to take as much as nature will give them,—even a siesta after meals may be allowed, as sleep is for them the best restorative, and an anti-spasmodic of heroic force. In obstinate sleeplessness depending upon nervous irritation, a warm bath, before bedtime, is often as serviceable at the change of life as in infancy. More than once, in cases of great insomnia, I have had the patient wrapped in cotton wool, from head to foot, which gave comfort, warmth, and sound sleep.

With regard to the more serious mental disorders, which cannot be altogether separated from insanity, it will often devolve on the medical man to explain to the patient's relatives, that the system may be, for a time, unhinged; that harsh conduct would aggravate the already too distressing symptoms; and that it would be cruel to consider as confirmed insanity the strangeness of temper, the fitfulness of spirits, the perversion of character, which, after a few weeks or months of treatment, may considerably abate or entirely disappear. The temporary avoidance of those to whom an aversion may be held, following the instinct of the patient so far as frequenting society is concerned, encouraging any inclination for particular occupations, are all points of great importance; for manual labour cuts off the heart from its too engrossing objects of delight, and if it is so useful, in cases of confirmed insanity, how much more so it must be to diminish the tendency to climacteric mental derangement.

Being able now, as years accumulate, to appeal to a large experience of the dangers that attend puberty and the change of life, I am at a loss to know which crisis most requires

judicious management, or the intelligent sympathy of friends,
and the habitual support of some stronger mind, in whom
faith is placed.   It is certainly surprising that a stout lady,
in excellent bodily health, eating heartily, and taking half a
bottle of wine a day, should be put out by trifles, be low-
spirited and fretful on the slightest mental exertion, and be
laid up with a severe attack of cerebral neuralgia if the
exertion be persevered in ; but such cases are not unfrequently
met with, and they may last one, two, or more years, after
which the patient gradually recovers her right tone of mind.
Premature efforts to resume the usual active habits of life
increase the frequency and severity of these attacks of cerebral
neuralgia, and compromise the chance of ultimate recovery.
I know a lady, of a good constitution, though somewhat
nervous all through life, and who, just as menstruation was
ceasing, lost a dear sister, and then a dearer husband.   Her
mental energy and power of action were completely prostrated,
but as she continued to look well and eat heartily, those
around her could not understand that she was unable to fulfil
the duties of her station, and repeatedly urged her to exertion;
although every fresh effort was followed by a breakdown.
More frequent attacks of cerebral neuralgia made her silly for
a time, and she never quite recovered her former healthy
state of mind.   Attention to the regular action of the bowels,
a purgative every two or three weeks, mild sedatives, change
and travelling, are the best means that I can devise to tide
over the mental dangers of this period.

Travelling is a great strengthener of the nervous system, for
it places the patient in entirely new circumstances, every one
of which makes a fresh call on her attention, solicits her
interests, captivates her faculties, and completely leads her
from trains of thought, to which she had been long enchained.
In addition to this, the exercise of various kinds which is
willingly taken gives increased vigour to the muscles of the
body, and therefore to the brain itself; the mind is often
consoled by the probability of recovering health by an agreeable residence at some watering-place, it feels under the

guidance of medical authority, and resumes peace and tranquillity. Indeed, all that has been said in praise of baths, exercise, and the advantage of a light-hearted cultivation of the pleasures of nature, rather than of those of society, by women at the change of life, will clearly point out how their combination may be made serviceable for the cure and prevention of disease at this epoch. Those blessed with ample fortunes can easily avail themselves of this provision safely to get over this critical period, and remodel their constitution so that it may endure for many years, and I have elsewhere stated the great advantages that may be obtained from the judicious use of mineral waters.

Mental and Moral Hygiene.—I have already said that during the change of life, the nervous system is so unhinged that the management of the mental and moral faculties often taxes the ingenuity of the medical confidant. The study of the patient's character will teach him what occupation or pursuit is most likely to engross her mind, and effectually replace those of former times. If he be not prepared to take the part of divine, moralist, and philosopher, without ceasing to be a physician, his medicines will often be of little use. It cannot be wondered at, that the full conviction that age has stamped them with its first irrevocable seal, should cast a gloom over the imagination ; but a well-trained mind will soon adapt itself to a new state, and take some comfort from the knowledge that this epoch proclaims an immunity from the perils of child-bearing, and the tedious annoyances of a monthly restraint.

The natural good sense of many will show them how the notion that after this period little remains to console women for the anxieties and troubles of life, is a *pagan* idea, suited to the position allotted to women in Greece and Rome, where they were seldom considered worth more than to amuse men and to bring forth children ; a condition inapplicable to women after their social emancipation by the doctrines of Christianity. They should be shown that the importance of their position after the great change, may be inferred by the

length of life allotted to them after its occurrence, and the
singular immunity from disease which is generally observed
after that period.   They should be reminded that many intimate sources of pleasure are attached to every age, although
it would be unfair to ask for one period the pleasures allotted
to another.

> " Qui n'a pas l'esprit de son âge,
> De son âge a tout le malheur."

Besides this vast improvement in health, it must not be
thought that the change of life implies the loss of all
personal attractions.   There is a beauty about childhood
that seldom makes a vain appeal for fostering care ; the
beauty of youth fascinates ; that of mature age excites
admiration ; but in many women there is for long after
the change of life an autumnal majesty so blended with
amiability, that it charms all who approach within its
magic circle.   To those fired with ambition, it may be
safely said, that the home government of society, from
the highest to the lowest of our social strata, offers a wide
field of employment to women after this period of life.
Many never think of cultivating their minds until they find
their influence fading with their charms, and then set about
acquiring a less perishable empire, and employ this period of
freedom in literary pursuits.   Others govern with discretion
that circle of society, limited or extensive, in which they
have been placed; becoming the guides, the supports, and
mainstays of both sexes in the difficulties of life.   Indeed, it
would not be too much to say, that the discordant elements
of society can never be blended without the authority willingly conceded to the combined influence of age and sex.   It
is a matter of history how society has been modified by the
drawing-rooms of Mme. Lambert, Mme. de Tencin, Mme. du
Deffand.   Those acquainted with French society in our own
time will remember the reign of the late Mme. Recamier over
a large circle of talented friends ; and during my residence in
Paris, I have myself frequently witnessed the benign in-

fluence of Mme. Swetchine* prompting the eminent men who gathered around her to what was great and noble; guiding many women in the difficult paths of life; healing wounds caused by inexperience; and making many bless her still, for the happiness they now enjoy. This brings me to the noblest motive to be offered to the laudable ambition of women—that of doing the greatest amount of good, to the greatest number of their fellow-creatures. Time dulls the eye, robs the cheek of its bloom, delves furrows in the forehead, but cannot quell the seraphic fire burning in the heart of women, prompting them to deeds of charity, and to heal the deep and ever-breaking-out wounds which afflict society. Those who have attained their sunset without having been granted the anxious though desirable vicissitudes of wedded life, even if destitute of relatives, or unfortunate in friendship, may still find in the various forms of unmerited affliction which fill our country cottages, or the hovels of our populous cities, that whereon to expend a warmth of feeling, an energy of self-sacrifice, which the sophisticated state of society has not permitted to flow into their natural and more grateful channels. Why then should women, sensible in everything else, be sometimes unable to accept their new position, and, instead of kind and charitable, become, at this period, peevish, harsh, and dismal, viewing everything through a jaundiced veil?

The most distressing appeals to medical sympathy, are made by those who, when unnerved by the change of life, find themselves alone in the world, bereft, when most needed, of the solace of filial piety, or the supporting sympathy of conjugal affection. One can only at first, respond to such appeals, with a sympathizing look and a silent pressure of the hand; but should tears burst their bounds, lightening the

---

* Since these lines were written in the Second Edition, Mme. Swetchine is dead; Cte. de Falloux has published her works, which have been translated into English, and her " Pensées Choisies" have been beautifully illuminated by my friend Miss Simpson.

suffering spirit of half its load, sweeping away sorrow, dis-
quietude, and doubt, then it may be hinted that time steals
even sorrows from the heart, doubtless because they are
sweeter than joy, and that after a brief period nothing will
remain but the clear, calm, and grateful remembrance of
past goodness, where it was once thought that impassioned
love and ever-devoted tenderness must be eternal.   Here
again the best mode of affording relief, is to discover some
kind of occupation capable of engrossing the sufferer's
attention, such as music, gardening, the education of a
relative or of an adopted child, or the management of a
school, or some other charity.   The continued friction of
social duties will, in time, rub off the asperities of character,
and restore peace and tranquillity to the troubled spirit.
Every effort should be made, in such a case, to prevent
brooding and self-absorption, for the mind may gaze so long
upon itself and its inner workings that the moral vision may
become affected with the same kind of disorder which befals
the bodily eye when fixed too long on one colour ; surrounding
objects then lose their real appearance, to shine only with
unnatural tints.

PART III.

SPECIAL PATHOLOGY

OF THE

CHANGE OF LIFE.

CHAPTER VI.

DISEASES OF THE GANGLIONIC NERVOUS SYSTEM.

TABLE XXIV.

*Liability to Diseases of the Ganglionic Nervous System, in
500 Women at the Change of Life.*

Epigastric faintness and sinking	220
Epigastric pain or anomalous sensations .	49
Fainting or leipothymia.	25
Fainting off, for the first time in life . .	3
Prolonged and intense debility	41
Monthly depression of strength	1
Chlorosis	40
Palpitations	16
Aortic pulsation	2
Hysterical asthma	7
Monthly hysterical asthma.	2
	406

Thus 406 instances of ganglionic suffering were divided
amongst the 500 women who suffered more or less from diseases
of the ganglionic system; many of them suffering in more ways
than one. At the same time the liability to these affections
decreases as women advance in age; epigastric pain is then
seldom intense, but there frequently remains a liability to
fainting, and to sinking sensations at the pit of the stomach.

The volume of the brain, its complicated and regular

structure, show it to be contrived for important ends, and
although the relation existing bstween the intellectual faculties
and the structure of the brain cannot be comprehended, it
must be admitted. It is far different with respect to the
ganglionic nervous system, and indeed it is incomprehensible
that so much vital force, for good or for evil, should be centralized in little irregular lumps of nervous matter, and in
sundry tangled skeins of nerves, the geography of which, like
that of the polar regions, is differently mapped out by successive observers; but though incomprehensible, it is no less
certain that these knots of nervous matter, and these tangled
skeins of nerves, control the blood-vessels, and are indissolubly
connected with the supreme power which guides the processes
of healthy or diseased nutrition. This is proved by the
writings of Winslow, Bichat, Reil, Wilson, Brachet, Philip,
Broussais, and Lobstein, besides the recent writers quoted
in the following pages. Whether, with Bichat, anatomists
look upon the ganglionic nervous system as independent, or
with Haller, as an offshoot of the cerebro-spinal system, the
ganglia contain every variety of nerve tissue, and must still
be considered as little brains, each having its special range
of power, deriving sensitiveness from the sensory nerves,
which proceed to every ganglion, by means of which nerves
the brain becomes cognisant of morbid ganglionic impressions.
With Müller, I consider the ganglia to be the source of the
energies of the sympathetic nerves, and the fountains from
which the ganglionic nerves draw the constant, gradual
galvanoid current by which the capillaries throughout the
frame are enabled to do their work. It, moreover, appears
that each separate ganglion sends its contingent of nervous
influence to the central ganglia, and that the force with which
the ganglionic nervous system is endowed, is centralized
in the epigastric region as the intellectual faculties are in
the brain. Discordant as medical theories generally are,
it is singular how often the importance of considering vital
force as centralized in this epigastric centre has been pro-
minently asserted; Galen and Fernellius called it the prin-

cipal lever of the human forces; Van Helmont there placed his *Archæus*, Wrisberg and Lobstein treated of it as *the cerebrum abdominale;* Hunter called it the sensitive centre, and the centre of sympathies; and Bichat, Broussais, &c. considered it the prime conductor of nervous influence. The importance of this region as a centre of power, is even shown by the erroneous theories which made some medical men place the seat of sensibility in the diaphragm, and by the popular belief that the human passions are centred in the præcordia, whereas they merely react upon it, as stimulants when the passions are exhilarating, or as depressants when they are of a contrary nature.

Nevertheless, the ganglionic nervous system, in a pathological point of view, is looked upon as a *terra incognita;* and its diseases are neglected or very incompletely considered in systematic works on pathology, and are scarcely better treated in books on nervous diseases—a neglect partly due to the wild theories that have been broached respecting the ganglionic nervous system by fanciful writers. Thus, notwithstanding Lobstein's classic work, Georget, writing in 1836, affirmed, that nothing is known about the diseases of the solar plexus, or the ganglionic system of nerves, and a little later, Sir H. Holland stated "that the ganglionic system, and the various nerves of organic life, are still only partially known to us in their proper actions, and yet more obscurely in their intricate connexions with the nervous powers of animal life;" and that, "a less definite influence of the system of organic life, as one of the causes of exhaustion, is the only morbid liability of this all-pervading nervous system." It is not surprising, however, that the pathology of the ganglionic system should have been hitherto imperfectly considered, since its physiology is full of desiderata; but the experiments performed by Claude Bernard, and the inference drawn from them by Brown-Séquard, that the blood-vessels were paralysed by the section of their ganglionic or vaso-motor nerves, have given a physiological basis to the study of the diseases of the ganglionic nervous system. These

K

experiments establish the fact that the partial paralysis of the ganglionic nerves leads to the congestion or to the inflammation of the tissues within the area of their distribution, and the controlling influence of the semilunar ganglia over the blood-vessels of the abdominal viscera is shown by the experiments of Samuel, who found their extirpation in dogs, cats, and rabbits to give rise to an extraordinary amount of hyperæmia of the intestinal mucous membrane.

Besides controlling circulation, it would seem as if the ganglionic system furnished to the human frame a nervous influence that plays the part of steam in the steam-engine, the unconscious nerve force giving to the cerebro-spinal system a power, of which the human mind has the full consciousness. At all events, any one who will take a comprehensive view of the various stages of the reproductive process in women, will be struck with the fact that frequent prostration of strength is a predominant symptom even when women are healthy. At every recurring menstrual period, at the cessation of menstruation, after connexion, parturition, and during lactation, there is felt, more or less, a loss of energy ;—so it seems as if woman could not pass through any of the stages of that function which serves to communicate life, without the momentary loss of some portion of her own vital energy, reminding one of those animals who die when once they have transmitted life to others.

Et quasi cursores vitæ lampada tradunt.

A glance at the preceding table will show that debility of variable intensity is the constant characteristic of all the complaints which I attribute to the diseased action of the ganglionic system, and Dr. Handfield Jones,* who has seriously taken in hand the study of ganglionic affections, has well pointed out that, however painful may be cerebro-spinal neuralgia, it does not induce syncope or leipothymia,

---

* " Functional Nervous Diseases."

which are apt to accompany the irritation of any of the
sympathetic plexuses. Debility is, of course, often met
without our being able to trace its origin to any particular
part of the body; but debility is so intense when disease of
the central ganglia is evident, and so constant an attendant
upon every form of gangliopathy, that we may consider debility as its main symptom. The same remark applies to
chlorosis, and still more forcibly to leipothymia and syncope.

Gangliopathy.—I give this name to a condition in which
more or less debility is associated with paralysis, hyperæsthesia or dysæsthesia of the solar plexus and the central
ganglia of the sympathetic system. Gangliopathy has often
been described as cardialgia, gastralgia, and gastrodynia, but
these names should be restricted to neuralgic states of the
stomach. Gangliopathy has been written of, as neurosis of
the vagus, as leipothymia, as sinking dyspepsia, as hysterical
asthma, and one of my cases had been called by other practitioners *a nervous affection of the diaphragm.* Women frequently
complain of *chest pains,* or of *spasms, inward spasms,* or *inward
hysterics.*

The effects of a moderate blow to the pit of the stomach
will explain what I mean by gangliopathy being attended by
paralysis. For instances of perverted ganglionic action, I
refer practitioners to the most prominent symptoms of over-
lactation; but, as I am drawing attention to a disease that
has been to a great extent overlooked, I shall begin by giving
cases that illustrate the various forms of gangliopathy, and
I prefer to use a term that localizes the disease without
specifying its nature, because I think the present state of
science seldom warrants a more definite expression.

CASE 15.—*Ganglionic hyperæsthesia.*—Miss C. was forty-
eight, tall, stout, with dark hair and a flushed face. The
menstrual flow came regularly from thirteen to forty-seven,
but afterwards irregularly, being often a mere show. This
patient was never nervous or hysterical; she complains of
pain at the pit of the stomach, which first appeared when
the menstrual flow became irregular, and says that now she is

K 2

never without uneasy sensations at the epigastric region,
which do not generally interfere with the current occupations of the day, though often paroxysms of acute pain
occur, especially at night, when they suddenly awaken her
from a sound sleep. The pain then experienced is described
as a "tearing pain," and after it has lasted from ten to
twenty minutes, a ropy mucus comes from the mouth, by
expuition, without eructations. When the intensity of the
pain has abated, the patient lies prostrate for hours, conscious, but incapable of exertion. Sometimes she faints
away, and after a bad attack was forced to keep her bed a
day or two. During the last six months, flushes and perspirations have been abundant. The tongue was clean, digestion good, and no trace of tumour at the pit of the stomach.
I recommended 6 oz. of blood to be taken from the arm,
2 tablespoonfuls of a comp. camph. mixture before, and
10 grs. of carb. of soda after meals ; 2 comp. col. pills and
10 grs. of Dover's powder on alternate nights, and a mustard
or a linseed poultice to be applied to the pit of the stomach
every night. The camphorated mixture that I often give in
similar cases is composed of 3 drachms of tincture of castor,
6 drachms of tincture of hyoscyamus, and 5 oz. of camphor
julep. After continuing this for a month, the paroxysms
came only once a week, instead of almost every night ; I
then ordered a warm bath to be taken for two hours every
other night, just before going to bed ; alternate belladonna
and opium plasters, changing them every week, and a scruple
of sulphur once a day. This was persisted in for six weeks,
and then left off, because there had been no paroxysms for
ten days. When the patient left town, I advised her to take
the mixture now and then, as well as the pills and the comp.
sulphur powder, and to have 2 or 3 oz. of blood withdrawn
from the arm at intervals of three and six months. This
case seems to me best accounted for, by admitting a neuralgic affection of the ganglionic nervous centre, for the
stomach performed all its functions healthily, and there was
no sign of cerebral disorder, neither did this affection obscure

the comprehension of its true nature by awakening other nervous disorders. It caused no hysteria, no pseudo-narcotism, not even headache. This neuralgia of the ganglionic centre was well characterized by the paroxysmal character of the pain, and by the state of exhaustion and faintness it determined.

CASE 16.—*Ganglionic hyperæsthesia.*—Mrs. K., aged forty-two, consulted me, July 5th, 1855. This lady is stout, of an average height, with dark hair and eyes, a swarthy complexion, and has all her life been subject to bilious complaints. The menstrual flow first appeared, with frequent fits of fainting, at twelve. It continued regular, and was often accompanied by brow-ague. She married at twenty-seven, but never conceived. In 1848, after having been a year at Ceylon, she was first taken with excruciating pains in the dorsal region of the spine, and in the abdomen, which pains always came on at night. She improved under a mercurial treatment, had the Bombay fever the following year, but came home recruited by the long sea voyage. Four years after her return, the menstrual flow became irregular, and now she never passes more than " a little green water" for one day. Since the menstrual flow became irregular, she has been always ailing; was once under Mr. Keate for an hysterical affection of the shoulder-joint and paralysis of the arm, and has often had attacks similar to those she had at Ceylon. She stated, that every night she was awoke by an acute, " gnawing, hot pain" about the fifth dorsal vertebra. This pain encircled the right side, reached the pit of the stomach, where it centred and kept tearing and gnawing her for about two hours. The patient very graphically described the sudden coming of the attack—" as if the thing called pain played a thumping overture upon her." She has learned to relieve these attacks by laudanum; in their absence, she is very often troubled with a burning pain at the pit of the stomach; and of late, when this was intense, she has often fainted several times a day. This lady, though stout and healthy-looking, is endowed with great nervous irritability, in-

creased by anxiety for a very nervous husband. She feels
every change in the weather, and suffers acutely when there
is thunder in the air. She is often heavy, stupid, drowsy,
and forgetful : very low-spirited, thinks she is going out of
her mind, and is often tempted to commit suicide. For the
last two years she has been much troubled with flushes, but
without perspirations; and circulation is inactive, for her
fingers are cold and blue, even on a warm summer's day.
The pulse is small, the tongue slightly furred. July 5th.—
I prescribed the comp. camph. mixture before meals, carbonate of soda after ; 6 grs. of blue pill, and 2 of ext. of hyoscyamus every other night, 10 grs. of Dover's powder every
night, a scruple of the sulphur and borax powder once
a day, and warm baths twice a week. July 18th.—Decided
improvement followed the sound sleep which came when the
treatment had been followed for a few days; the paroxysms
of neuralgia were no longer periodical, and she had only had
two since the 10th. I continued the same treatment, but
also ordered two belladonna plasters, one to the epigastrium,
the other to the painful part of the spine, which were to be
renewed every week. Aug. 15th.—The patient being very
bilious, I ordered 3 grs. of calomel every other night, and a
black draught in the morning ; the mixture, the soda, and
the Dover's powders were continued. She was salivated by
two of the pills. Sept. 19th.—The paroxysms of pain returned
again every night ; so I ordered the sixth of a grain of acetate
of morphia to be taken in a mixture, every two hours, until
sleep was induced. This procured fourteen hours of sound
sleep, and the attacks only occurred at long intervals. When
I last saw the patient, I advised her to take the c. camph.
mixture and the Dover's powder occasionally, to take, every
night, half a drachm of equal parts of sulphur and magnesia, and to induce sleep by acetate of morphia whenever
an attack was imminent. One point to be remarked in
this case is, that the frequent flushes were dry and burning, whereas gentle perspirations coincided with the improvement in the symptoms, and their persistence affords the

greatest chance of immunity from worse symptoms. This patient had been fearfully nervous all her life without ever having presented any decided form of hysteria, so varied are the states of nervous suffering, for many of which there is no name.

In other cases, it is not from intense pain that the patient suffers, but from annoying and singular ganglionic sensations, and as these do not depend upon any organic affection of the heart or the aorta, and coincide with other nervous symptoms of cessation, it seems fair to ascribe them to some morbid condition of the ganglionic nervous centre.

CASE 17.—*Ganglionic dysæsthesia.*—Sarah B., tall, stout, and healthy-looking, with brown hair, and hazel eyes, was forty-seven when she came to the Paddington Dispensary, Sept. 8th, 1849. The menstrual flow first appeared at seventeen, was always regular, and accompanied by pseudo-narcotism. She married at twenty-five, had two children, and the menstrual flow left suddenly, without known cause, at forty-four. Since then, she has been entirely free from lumbo-abdominal pains, but has suffered much from nervous symptoms. There has been no headache, but a heavy, stupid feeling in the head, with drowsiness in the day, after sleeping well at night, and a forgetfulness of familiar things. She has also been nervous, desponding, and low-spirited; often shedding tears, and complaining of strange sensations in the throat. Ever since cessation, she has been distressed by a fluttering at the pit of the stomach, "as if something were perpetually swinging within her." It becomes worse after meals, generally abates when she lies down, is seldom felt when in bed, but begins so soon as she rises. When turning the corner of a street, this sensation makes her feel afraid of losing her centre of gravity and overbalancing herself; and when she has it in bed, she feels "as if a tub were rolling to and fro within her," and then "the head goes too," as "if something rose from the pit of the stomach to the head, making it feel giddy and bewildered." Since cessation, she has been troubled by burning flushes, without perspirations ; and there is sometimes a good deal of pudendal irritation. There was no organic disease of the heart, aortic pulsation, or obstinate dyspeptic condition, to explain this singular symptom ; but it would be illogical to deny the patient's statement because her sensations could not be explained ; she has consulted many practitioners, and had been told "it was all nonsense." Sept. 8th.—I ordered the comp. camph. mixture before meals, and on going to bed ; carb. of soda after meals ; a large opium plaster to the pit of the stomach ; and a small teaspoonful of sulphur and carbonate of magnesia every night. September 15th.—She was better; I ordered a saturnine lotion for the pudendal irritation, and 10 grs. of Dover's powder to be taken every night. Oct. 6th.—Instead of perspirations, a papular eruption has appeared on the shoulders. She feels rather worse than better, but the same remedies were ordered, with the addition of 2 comp. col. pills to be taken occasionally. Oct. 20th.—All the cerebral symptoms have vanished. She is much better, and can now bustle about ; but the swinging sensation in the epigastric region still remains. The improvement coincided with gentle, well-sustained perspirations. I ordered the mixture and soda as before, but discontinued the sulphur and Dover's powders ; prescribing, instead, sulphur, 2 oz.; borax, 1 oz.; Dover's powder, 1 drachm, 2 scruples of the powder to be taken in a little milk at night. A blister was ordered to the pit of the stomach. Nov. 6th. —She looks cool and comfortable, much stronger, and quite like another person. Though less troubled by the swinging sensation, it still exists. The blister did no good, so I ordered a rotation of belladonna and opium plasters, each to be worn a week on the epigastric region, and the mixture and comp. sulphur powders to be continued. Nov. 23.— The patient was discharged cured ; she felt comfortable and happy at having lost the epigastric sensations.

CASE 18.—*Ganglionic dysæsthesia.*—Sarah J., an average-sized woman, with dark-brown hair, grey eyes, and a semi-chlorotic complexion, came to the Farringdon Dispensary,

February 25th, 1853, being then forty-six, and unmarried. The menstrual flow appeared at sixteen, and continued regular, without any nervous or other symptoms, except that, for two or three days, it was almost always preceded by the eructation of acid mucus. Flooding came on seven months ago, without any known cause, and the menstrual flow has since appeared every six or eight weeks, varying in quantity, and being accompanied by abdominal bearing-down pains, and severe headaches, although she has brought up much less acid mucus. She has suffered from lowness of spirits, involuntary tears, choking sensations, gasping for breath, and a sensation of fluttering at the pit of the stomach, which she compares to a steam-engine pumping up something from the stomach to the head, inducing headache and giddiness. These sensations are not always felt, coming on when she lies down, so she sleeps propped up in bed. The attack seldom lasts more than half an hour, but the night is restless, and the following day she sleeps a good deal in her armchair. I could detect no disease of the heart, no aortic pulsation, slight signs of gastric disorder, and this patient had no flushes or perspirations. I ordered the comp. camph. mixt. before, and the carb. of soda after meals, 10 grs. of Dover's powder at night, a scruple of the sulphur and borax powder twice a day, a belladonna plaster to the pit of the stomach, and 2 comp. col. pills to be taken occasionally. This treatment was continued three months, and completely cured the symptoms. The patient then became regular, and continued so for six months, when the menstrual flow ceased suddenly. She was then much troubled by great debility, headache, giddiness : and for the last year, has been subject to flushes and perspirations, but the epigastric annoyance has not returned.

CASE 19.—*Ganglionic paralysis.*—A widow lady, aged fifty-four, was regular until forty-one, when the menstrual flow ceased suddenly from being frightened at a revolutionary tumult at Athens. She then became subject to headache and bilious vomiting, which abated on her return to England in 1851. In 1855, the habit of vomiting completely ceased, but was succeeded by an overpowering sensation of faintness at the pit of the stomach. This sensation so exhausted her strength that she was confined for days to her bed. This complaint had been called "a nervous affection of the diaphragm," and aperients, tonics, and stimulants, with blisters to the pit of the stomach, were tried ineffectually until combined with sedatives, given as in the previous cases, which relieved this patient, notwithstanding ever-recurring domestic annoyances. In 1859 she wrote to say that she had recovered her health, and was able to take very active exercise.

The same treatment was successful, in similar cases, which occurred in Dispensary practice, and if these anomalous symptoms only occurred in those who have plenty of time to watch the coming of each new pain, and to nurse it until it has taken root in the nervous system, I should lay little stress on such cases; but when they occur in the sturdiest of Eve's daughters, hard-working women without imagination, I look upon the symptoms as indicative of local disease, and not of mental delusion. A strong-looking washerwoman, aged forty-seven, mentioned that, after the menstrual flow had ceased for six months she was, for the first time in her life, obliged to give up work, because slight exertion caused great pain at the pit of the stomach, with fainting fits often prolonged for three quarters of an hour. Mrs. W. ceased at fifty-six, and is now fifty-nine. After mental excitement, during the last year, she has been subject to aching pains at the pit of the stomach. She sleeps well until five A.M., and then arise epigastric sensations, as if something were turning in the stomach. This is followed by palpitations. She has no dyspepsia, but feels nervous, frightened, cannot cry, and, like the Wandering Jew,—" she must march on." Belladonna plasters entirely removed the epigastric sensations in a month. Phœbe M., aged fifty-six, has been unusually weak for the last five years, ever since cessation. She says she is troubled with "inward hysterics," and "faint heats." There is no aortic pulsation to be detected, but the least thing is said to bring on a violent abdominal beating, and causes her to pass two or three liquid stools, with blood from the bowels, or from piles. At times she loses her sight, and cannot sleep. Flushes are generally preceded by heat, pain, or faintness, at the epigastric region, and the sensations have been described by patients as "a ball of fire," or "gnawing, tearing pain," relieved by pressure, and prompting the sufferer to turn on her stomach to press the painful region. Whoever has seen patients suffering from lead colic will recognise similar symptoms.

Many patients, similarly afflicted, complain of the epigastric region being "all of a boil," and that any sudden impression sets the stomach "all of a flutter," with great subsequent weakness. Dr. Brown, of Newcastle, has met with a case of emotional insanity, in which the patient had the sensation of a bright flash of lightning in the epigastrium, and three other lunatics complained of darkness in the same region.

In the following cases the difficulty of breathing from a reflex spasmodic contraction of the diaphragm and intercostal muscles was a predominant symptom. The following case was published by Dr. Marrotte (*Gaz. des Hôp.* 1854), under the title of "*Spasme simultané de la glotte et du diaphragme.*"

CASE 20.—*Gaygliopathy and hysterical asthma.*—A woman, aged forty-four, entered the Hospital of "la Pitié." The menstrual flow came at seventeen, with disturbance of health, and at twenty-three she had frequent menstrual irregularities; became nervous, had agonizing sensations at the epigastrium, with globus hystericus, and a spasmodic contraction of the diaphragm, continuing so long that her friends feared asphyxia. In these attacks, which occurred about three or four times a week, she heard what was said, but could not speak. Cessation took place at forty-two, and the nervousness increased, for she laughed or cried at the least thing. The day the patient entered the hospital the house-surgeon saw her in one of these attacks. Pain at the pit of the stomach, and a sensation of stricture at the basis of the chest were first felt, then respiration was suspended, after which followed great efforts to inspire and expire, both being separated by variable intervals, and accompanied by laryngeal whistling. The noise was strongest during inspiration, which was attended by an exaggerated dilatation of the chest, and great efforts to swallow an imaginary bolus. Several of these attacks occurred, lasting from half-a-minute to three minutes, and were followed by pain at the pit of the stomach, by fainting and exhaustion. The attacks seemed to abate on pressing the pneumogastric nerves at the inferior edge of the thyroid cartilage.

Although there were no ganglionic symptoms in the following case, it much resembles the previous one.

CASE 21.—*Hysterical asthma.*—Miss O., tall, stout, of a sanguine temperament, was fifty-seven years of age. The menstrual flow came at thirteen, and was regular until fifty, when it ceased gradually. Until then she never had any serious illness, and no marked hysterical symptoms, though she was always very excitable. Just before cessation she had the shingles, and soon after erysipelas, but with the exception of flushes and gentle perspirations she remained well for two years. Five years ago, Miss O. was, for the first time, and without known cause, seized with the nervous affection for which she consulted me. Without being sick, she goes through the pantomime of sickness; she gasps for breath, but retains consciousness, and does not suffer from positive pain, but from the anguish of not being able to breathe. The attack lasts from three to ten minutes, and is often preceded for several days by a spasmodic affection of the jaws, "as if she could not put the teeth in the right place." A neuralgic flash of pain across the forehead and heavy sweats, are its immediate precursors, and it leaves her in such a state of exhaustion, that she has often kept her bed for several days. These attacks used to occur at short intervals, but lately there has been a longer period between

them. Worry brought one on a month ago. She has no
epigastric pain during the attack, or at any other time. For
the last three years there has been a red sediment in the
urine, and an unusually frequent desire of passing it. She
has no leucorrhœa, but for two years she has been frequently
troubled by great irritation of the labia. Hard lumps are said
to arise, and, after remaining a few days, to disappear, without
suppurating, but sometimes end by a pimple, and she has suffered sometimes from piles. Miss O. is evidently very nervous,
and, although grown stouter, she is very weak, and as her
attacks were becoming less and less frequent, no active treatment was required. I recommended the comp. camph. mixture to be taken, off and on, for several months; twenty
drops of liquor potassæ in a little water, after meals, and the
sulphur and borax at night. Warm baths were also suggested, particularly when there was pudendal irritation, and
citrate of quinine and iron. This patient enjoyed such
excellent health until cessation, that I attribute all the subsequent illness, the nervous attacks, the erysipelas, and the
pudendal irritation, to the organic changes set on foot by the
turn of life, and some years afterwards, I learnt that Miss O.
had only felt slight returns of the same symptoms.

Voisin also relates, that a case of hysterical hemiplegia
was sometimes attended by instinctive and prodigious efforts
to breathe, with emission of a singular noise from the glottis.
Dr. Marrotte considered his case analogous to the laryngismus
stridulus of children; but when the subject was discussed
at the *Société de Chirurgie de Paris*, Dr. Beau judiciously
observed, that there was no analogy between the two affections. At p. 84 of my work on " Uterine and Ovarian Inflammation," I have given the case of a lady, aged thirty-three,
whom I have frequently seen similarly attacked; there was
no laryngeal spasm, but absence of all respiratory movements
for five or six minutes, as she lay immovable, and with open
mouth. The gradual solution of the spasmodic contraction
of the diaphragm could be measured by the increased length
of respiration, a full respiration bringing back consciousness.

Sir B. Brodie mentions the case of a young lady, in whom the slightest pressure on the epigastric region would bring on paroxysms of suffocation from spasmodic contraction of the walls of the chest. In a diminished degree, the symptoms frequently attend an ordinary hysterical attack, though they may occur in an aggravated form in women suffering little from the ordinary forms of hysteria, and who present no signs of emphysema, or of heart or lung disease. In my six other cases of hysterical asthma at the change of life, the symptoms were less marked than in that of Miss O.; but those described by Dr. Marrotte were reputed not uncommon by Dr. Beau, who mentions having seen a woman, suddenly taken with suffocation, and die in two minutes, without any discoverable cause. M. Bacchias, in his thesis, has given several similar cases.

Having described the minor, as well as the most severe forms of cardialgia,—having adduced examples of it, where the prominent symptom was pain or sensations almost baffling description, and complicated by various reflex neuralgias, it is well to inquire the opinions of pathologists, who have described the severe forms of the disease of the ganglionic centre. Hippocrates has noticed agonizing epigastric pain, accompanied by the impossibility of breathing, as having occurred at puberty. F. Hoffman, in describing cardialgia, insists on the intense pain at the pit of the stomach, the sense of anxiety, the difficulty of breathing, and the prostration of strength, opining with Barras, that the stomach was alone affected. Some of Schmidtman's cases of this disease are similar to mine, like that of a girl of sixteen, who, for several months before menstruation, suffered severely, at the pit of the stomach, for a few days every month. Barras adopted, I think erroneously, the term Gastralgia, for the ganglionic phenomena occurring at lactation, chlorosis, or pregnancy; and remarks, that his patients refer the sensations of weakness and prostration which overpower them, to the epigastric region. Joly and Georget also described these symptoms as Gastralgia, and noted its frequency in women during the reproductive period of life, or whenever the system is strongly reacted on by the reproductive apparatus. Louyer Villermay gives cases in which hysteria is accompanied by attacks of intense epigastric pain, exhaustion, syncope, or suffocation. Dr. Addison has not exaggerated the frequency of gangliopathy in those suffering from uterine affections, and Hufeland and Chambou state that, at the change of life, women are subject to suffocation and epigastric spasms. Dusourd, remarking on the same phenomena, says, "they simulate asthma," and adds, that "the affections and sensations at the pit of the stomach, at the change of life, baffle description."

Dr. Shearman—*Medical Times*, September 20th, 1856—bearing in mind the fact, already well established by Romberg, that neuralgia of the vagus nerve is indicated by patients becoming ravenous, who could previously take only the smallest quantity of food; and coupling this with Bernard's discovery, that irritation of the origin of the vagus caused sugar to form in the liver, to be removed by the kidneys, and appear in the urine, Dr. Shearman "concluded that the disease was one of the vagus nerve, in some parts of its tract, either centric, peripheral, or intermediate; of irritation in its early stages, and of palsy succeeding to that irritation." He designated it *sinking dyspepsia*, and treated it as he would the neurosis of any other centripetal nerve. If I be not deceived, Dr. Shearman has exaggerated the frequency of neurosis of the vagus, describing as such, those affections of the ganglia and cœliac plexus, which are more or less complicated by reflex neuralgia of the vagus nerve. The vagus is a bridge by which the central portions of the two nervous systems are placed in communication; uniting both nervous systems it resembles both, but principally the sympathetic, by its organization and the modes of its distribution. It is not a well isolated nerve, for it often anastomoses with the sympathetic, and helps to form the cœliac plexus, by nerves which are both numerous and voluminous. Thus anatomically united, I believe the affections of the vagus and of the epigastric ganglia generally coincide.

Causes of gangliopathy.—The ganglionic nervous system, particularly its epigastric centre, may be morbidly influenced, 1stly, by concussion ; 2ndly, by mental emotion ; 3rdly, by heat ; 4thly, by cold ; 5thly, by poisons.

1stly. A blow to the pit of the stomach does not kill by inflammation, but by a shock, so suddenly intense, that the laws of pugilism forbid " to hit under the belt." The blow may, however, be so graduated as to determine corresponding shades of neuralgia and of leipothymia. The recoil of a pistol, to the pit of the stomach of a lad aged thirteen, made him faint and lose consciousness for several minutes, emptied the stomach and rectum, and for fifteen minutes the heart's action was very feeble and the radial pulse could not be felt. The anguish of vomiting is partly due to the more or less intense concussion of the epigastric ganglia, which explains the intense debility it causes, and the not unfrequent deaths that occur when, as on board ship, it cannot be checked.

2ndly. Great sudden and depressing emotion may act on the ganglionic centre like a blow, setting up morbid action, lasting for twenty years, as in the last case of the next chapter, or for a shorter period, as in a dispensary patient, aged forty-seven, who was ceasing to menstruate, and was in good health when a son came home ill; and what with the fatigue from nursing and the fear that her son would commit suicide, she was suddenly seized with great epigastric pain, and a shaking of the abdomen so distressing that she was obliged to stop quiet, and support it with her hands. This symptom was most troublesome at night, and was made worse by exertion or worry. A flooding came on, and the patient gradually mended.

I have met with cases in which connexion acted as a depressing poison to the ganglionic nervous system, causing intense epigastric pain, unconsciousness, and great prostration, lasting for two or three days. There was nothing to explain this constitutional peculiarity except the fact of con-nexion being unattended by pleasure.

Besides the capability of being stunned by a blow or sud-den emotion, the great central ganglia are susceptible of receiving shocks from self-generated poisons spontaneously evolved in the system. " Miseros vidi ægrotos," says Lob-stein, "qui, vix somno dediti, subitò fuerunt expergefacti atque valido et quasi electrico ictu territi, ab epigastrio pro-ficiscente; crudele phenomenon, quod per plurimum mensium spatium duraverat." Other authors have noticed the same strange sensation, but independently of these rare occur-rences, it will be obvious to the pathologist that the gan-glionic centre may receive, from causes spontaneously arising in the frame, milder shocks, which determine the varied forms and degrees of ganglionic disturbance.

Emotions, varied feelings of pain or pleasure, past, present, or anticipated, are conceived in the brain, and sent, lightning-like, along the nervous cords to the various ganglia which stimulate the viscera to respond by appropriate action. This is admitted by everybody, but the converse is equally true that the viscera by their ganglia react on the central epigastric ganglia, and thereby on the brain, so as to suggest to this supreme organ the emotions whose purpose they subserve. What is obvious with regard to emotions, is very probable with regard to the physiological and pathological stimuli that the viscera transmit to each other and to the brain. In both sexes, there is more or less of paralysis and perturbed action of the central ganglia during convalescence, in con-sequence of the undue strain put on the ganglionic system to support the increased effort made to repair the effects of disease; in helminthiasis, when the expansions of ganglionic nerves are irritated by worms ; in agues, of which the ganglionic system is the prime motor; in hæmorrhoidal and other flows, when indiscreetly stopped by cold ; in chlorosis, hypochondriasis, and in the earliest stages of insanity, before the ganglionic, are cast into shade by the magnitude of cerebral symptoms. But the greatest and most frequent

L

cause of disturbance of the ganglionic centres, is the strong reaction of the reproductive organs : puberty, menstruation, pregnancy, lactation, and cessation, almost always cause slight forms of gangliopathy, and sometimes the severest, and may lead to insanity and to suicide. Woman suffers more than man, for her ganglionic nervous system is doubly taxed for self-nutrition and that of the race, and if she be susceptible of so often "tumbling to pieces," and of being again knitted firmly together, it is because her ganglionic nervous system has been endowed with extraordinary powers for good or for evil; but man does not escape, and I can confirm the accuracy of Schmidtman, who has paid so much attention to nervous affections, when he says, " so often as a young man consults me for cardialgia, I suspect onanism." Gangliopathy, under varied forms, is frequently observed in spermatorrhœa, and explains why some have committed suicide, and many have become hypochondriacs. The *debauché* and the *roué* are frequently at a loss for terms to express the annoyance of their sufferings at the pit of the stomach. These sensations in women at the change of life, may be the harbingers of others that obscure them, such as flushes, perspirations, diarrhœa, bloody stools, and hysterical symptoms.

3rdly. Dr. Morehead has noted the prominence of what I call ganglionic paralysis in what he describes as the cardiac variety of heatstroke. Denon relates that when marching through the burning plains of Upper Egypt, in the hottest part of the day, several men died of heat. " Nothing," says he, " can be more frightful than this death. Of a sudden comes intense sickness, then sinking and fainting that nothing can relieve, and death soon occurs." A French surgeon, describing the effects of the sirocco, on soldiers marching from Oran to Tlemcen, speaks of a kind of incubus weighing on the epigastric region, of debility almost reaching to syncope, of sudden bursts of heat in the face and intense headache. Similar phenomena, only less intense, may be caused by remaining too long, in Turkish Baths too much heated, as I know by my own experience of one I took at Damascus.

4thly. Dr. H. Jones has drawn attention to the paralysing influence of raw and intense cold, and to the manner in which it strikes at the pit of the stomach, yielding to brandy and warm epigastric clothing.

5thly. Poisons paralyse the ganglionic nervous system in various ways. *a.* Too much food or undigested food remaining in the stomach often gives rise to much of the ganglionic disturbance that characterizes nightmare, and sometimes to death in the aged, as suggested by Mr. Higginbotham. *b.* There is great similarity between some of the sufferings of hysteria and lead-poisoning. There is the same exquisite distress referred to the pit of the stomach, and the same instinct to seek relief by pressure to that part.

Symptoms and Diagnosis.—Debility underlies all other ganglionic affections, in the same way as nervous irritability underlies all cerebral diseases. Debility existed in a marked degree in all the cases enumerated above, although it was only marked, intense, and prolonged in forty-one instances, where it did not seem to depend upon any other cause than the change of life. Sometimes there was an overpowering sense of exhaustion pervading the whole system; thus forty-one women, of previously active habits, deplored being rendered helpless by intense and long-continued debility. A. M., a hard-working woman, says, "she hardly knows how to dress herself, and is often obliged to lie down to recover herself." M. W., a thin, delicate, red-haired woman, aged forty-five, two years ago got her feet wet at a menstrual period; the flow stopped; there were great abdominal pains, and a state of intense and long-continued prostration, without any other cause than the change of life. In others, the debility is so rooted in the system, that a full meal or stimulants will not dispel it. There is only a difference of degree, none of nature, between this thorough exhaustion of radical strength, and its slight manifestations, so frequent at this period, which seem to originate from the pit of the stomach.

This epigastric faintness occurred in 220 out of 500 cases, and is one of the most frequent of the minor torments to which women are subject. Pathologists have noted the severe cases of disease of the ganglionic centre, but they have failed to see the connexion between these and its milder forms, which occur habitually; whereas, between the severest of the 49 severe, and the slightest of the 220 mild cases, I only see a difference of degree, none in the nature of the complaint, which is as variable as the ever-varying shades of the same colour. The symptoms are described as "a sinking and faintness," "faint heats," "or a dull, sickly feeling at the pit of the stomach;" and the sensations are compared to those of "hunger, or to the craving of the dram-drinker for his wonted stimulus." Others feel "as if they could faint," as "if they wanted support," as "if they had no inside," as "if there was a vacuum." They will not allow the sensation to be called pain, but speak of its being so irritating and irksome that it is worse to bear than pain. They will frequently say that "all the complaint lies in the chest," but they point to the pit of the stomach, and have not a single symptom of disease of the lungs. There may be oppression and faintness, with or without violent pain at the pit of the stomach; sensations like "the fluttering of a bird," "the throbbing of an animal," or those "of internal rawness," "of gnawing and of tearing;" but these feelings do not depend upon gastric inflammation, since the stomach, even in some of the most distressing cases that I have related could digest any food; nor upon dyspepsia nor disease of the liver, and I therefore adopt Lobstein's opinion, that they indicate a state of suffering of the ganglionic nervous centre—"Hodie certissime evictum est, quod, tot numerosæ sensationes, quæ in epigastrio percipiuntur, neque ad musculos, neque ad vasa, neque ad organa gastrica sint referenda, sed unice ad plexum nervorum gangliosum, trunco cœliaco, insidentem, atque a Wrisbergio, summo cum jure *cerebrum abdominale* vocatum." At the same time it is evident that more or less ganglionic disturbance often accompanies disease of the stomach, liver, intestines, and of the womb. Romberg thinks " that the peculiar sense of fainting and annihilation accompanying the pain, is pathognomonic of cœliac neuralgia, and distinguishes it from neuralgia of the vagus," in which case there will be a sensation of stoppage in the œsophagus, of choking, or of scalding, rising from the stomach to the throat, with regurgitation of ropy mucus or vomiting ; copious perspirations or a greater flow of urine. Pressure often relieves, though a slight touch may sometimes increase the pain. Women instinctively unfasten their clothes, and on account of their liability to gangliopathy, many of my patients have left off stays. The debility entailed by a severe attack of this disease is beyond the conception of those who have not witnessed it. A good night's rest does not restore the usual strength, which seems drained in its fountain head, and is not recovered till the patient has had several nights of sound sleep. The intermittent character of the pain has been well indicated by Schmidtman, *Per intervalla vexat cardialgia et remittet intermittetque.* I have seen the worst paroxysms of gangliopathy come on in the midst of perfect health and without any apparent cause, but I have seldom seen them assume the tertian type, said by Dr. Shearman to be of frequent occurrence. Whether the presence of sugar in the urine is a pathognomonic sign of neurosis of the vagus, as stated by Dr. Shearman, is to me very doubtful, on account of the intimate connexion of the vagus with the cœliac ganglia, and this leads me to question whether the formation of sugar in the blood is not rather to be sought in that nervous system which is chiefly connected with nutrition ; but this is a fit subject for future researches. Admitting, with Romberg, that the sense of repletion after taking a small amount of food, and the absence of satiety after taking a large quantity, indicates hyperæsthesia of the gastric branches of the vagus, and that the accompanying sensations of burning or gnawing at the pit of the stomach, with overpowering debility, indicate cœliac neuralgia, the distinction has no practical bearings, as Romberg himself

admits. At puberty, during irregularly performed menstruation, pregnancy, lactation, or the change of life, the
pains at the pit of the stomach may be supposed to depend
on hyperæsthesia or paralysis of the solar ganglia and cœliac
plexus, particularly if the patient has been subject to causes
likely to produce nervous affections, or to diseases of the
sexual organs, if cerebro-spinal neuralgia co-exist and has
been intermittent, and if nervous symptoms be unexplained
by organic lesions of the stomach.

When debility is intense it passes by a gradually sliding
scale into the state called leipothymia by the older writers,
and may end in the total extinction of vital power.
Fainting is generally considered synonymous with syncope,
whereas *syncope* is the failure of the heart's action, *fainting*
is the sudden loss of ganglionic power, determining loss of
consciousness ; while *faintness* is the temporary depression of
ganglionic power, consciousness being unimpaired. It is
true that leipothymia and syncope may coincide, and produce each other; though in some of my severest cases of
gangliopathy, I have seen fainting occur, consciousness
diminished, and respiration imperfectly performed, while the
heart's action was undisturbed. Dr. Copeland has already
noticed this fact, and states, that in similar cases, he has
even sometimes found the pulse to indicate bleeding. By
fainting, I understand the " deliquium animi," or the
" defectio animi," of Celsus, the leipothymia of Sauvages,
whose definition is true to nature—" Subitanea et brevis
virium dejectio, superstite pulsus vigore, et cognoscendi
facultate." Syncope is rare at the change of life, whereas
a frequent liability to fainting, occurred in 25 out of 500
women. M. S., aged forty-seven, frequently fainted from
slight exertion at this epoch, but never swooned before
cessation. S. A., a strong-looking woman, aged forty, was
obliged to give up work six months after cessation, and
for the first time in her life, because exertion brought on
fainting, and irksome sensations of a load at the pit of the
stomach. She has several times fainted off for three quarters

of an hour. She is relieved by passing wind, is not dyspeptic, but nervous, and has globus hystericus. M. G. never fainted in her life until the dodging time, and then fainting fits occurred two or three times a week, for three years. A. L. frequently fainted before first and last menstruation, and during each menstrual epoch. C. S. had repeated fainting fits in the year following cessation. B. de Boismont has noticed fainting at the change of life, but he has not connected the comparatively rare cases of fainting with faintness, and the debility so often complained of; but I consider fainting to be linked, by insensible gradations, to the slightest sensation of epigastric faintness. The preceding considerations may throw light on some cases of sudden death, insufficiently explained by *post-mortem* examinations. Those who have written on ague, and more recently Dr. Handfield Jones, have noted the paralysing influence of paludal miasma over nervous power. Sir G. Blane relates how the Walcheren patients, when in full convalescence, would unaccountably drop down dead: and in the hospital of San Spirito, at Rome, I have seen a man, recovering from pernicious fever, expire suddenly, without any *post-mortem* appearances being found to explain the cause of death. Some of the cases of sudden death in puerperal women, already in full convalescence, seem to me susceptible of being accounted for by the previous observations. A tall, thin, pale-faced, flaxen-haired, inanimate lady had been confined a week, and was doing so well that her accoucheur had ceased his daily visits, when, on sitting up to take her usual food, she fell back, and suddenly expired. No clot in the pulmonary artery, nor anything else was found to explain this event, except a somewhat flabby state of the walls of the heart.

As women grow old, all the ganglionic nervous affections abate, except debility, which may persist, to a variable degree, until death occurs from the extinction of that amount of power allotted to the ganglionic system.

Chlorosis is another form of debility, of which the late

Dr. Merei affirmed that it is one of the most extensively diffused pathological conditions of the present age. It is, doubtless, most frequently met with in the years which follow puberty, but all should recognise its frequency at the critical periods of woman's reproductive life. Caseaux has pointed out its occurrence during pregnancy; Blaud admits it to be more common than is generally believed among adults; Dusourd has recognised it, at the change of life, and I have noticed it in 40 out of the 500 tabulated cases. Indeed, every one must have noticed the altered appearance that will sometimes take place at that period, the change of skin from a healthy hue, to the waxy, sallow tint, the blood-less lip, the pale white of the eye, with headache, pseudo-narcotism, and general debility. Leaving out of the question all cases in which anæmia was caused by flooding, it is well proved that puberty, pregnancy, lactation, and cessation may cause the relative proportions of the blood globules to fall low enough to produce chlorosis. This is doubtless a blood disease, but it originates in some peculiar loss of power of the ganglionic nervous system which moves the blood-vessels and vitalizes the blood. I have long believed, with Dr. Joly, of Paris, that chlorosis is a neuralgic affection of the ganglionic nervous system; and, while admitting with Dr. H. Jones, that in many cases of chlorosis occurring amongst the poorer classes in London, the action of malarious influences upon the ganglionic system is the first link in the chain of causation, such cases must not be severed from others equally frequent, wherein chlorosis arises under the healthiest circumstances and in the wealthiest families, showing that the ganglionic system may spontaneously de-velop the disease. At the change of life, as at puberty, the nervous energy of the ovaria is below par, or too great, and the epigastric centre, being morbidly influenced, is unable to promote the healthy performance of nutrition, so the blood is impoverished and the whole system suffers.

Therapeutics.—The cases related in this chapter show that gangliopathic affections are to be cured by sedatives, alkalies, by promoting the action of the skin, and by tonics; but I shall more particularly allude to these medications, and mention several remedies not yet alluded to by me. And to begin, I cannot sufficiently praise the utility of sedatives applied to the pit of the stomach.

The common consent of mankind, the convictions of many illustrious men, and the facts recorded in this chapter, go far to prove that, in the epigastric region, there is a real centre of nervous power, and I therefore seek to act upon it, by such remedies as seem likely to increase or diminish this power, or to regulate its disordered action. This should be done, if only to relieve the painful, distressing, and unaccountable sensations experienced by patients in this region; but being firmly convinced that this ganglionic nervous centre is in constant action and reaction on the brain, and having so often seen it derange more or less severely the mental faculties, at puberty, during pregnancy, puerperality, lactation, and cessation, I deem it imperative to apply remedies to this centre. Another important reason for trying, in every way, to relieve epigastric pain and anomalous sensations at the change of life, is, that women will not long endure these sufferings without seeking relief. If medicine does not relieve them, they will instinctively fly to stimulants, the poor to porter and gin, the rich to wine and brandy. Is it not then better to endeavour to alleviate these distressing symptoms by medicine, than to let women run the risk of becoming gradually addicted to deplorable habits?

The chapter on the general treatment of diseases at the change of life should be borne in mind, inasmuch as local applications can be of little utility without constitutional treatment. When the mild forms of cardialgia are presented, the epigastric uneasiness, the sinking and faintness, I first ascertain whether these sensations coexist with foul secretions requiring purgatives, before ordering my sedative mixture before meals, the alkali after meals; 3 grs. of blue pill and 2 of ext. of hyosc., every, or every other night; a mustard, or a hot linseed meal poultice, sprinkled with

coarsely-powdered camphor, and a teaspoonful of laudanum
every other night; dry cupping, as recommended by Galen,
and a piece of piline, or a camphor sachet, to be worn
during the day on the pit of the stomach. If the pains
continue, I prescribe a belladonna, or an opium plaster,
which is to be left on, and should it fail to relieve, I renew
it every fourth or fifth day, or I apply two; one to the pit
of the stomach, and the other to the opposite region of the
spinal column; or I order alternately, an opium or a belladonna plaster every fourth day. I often prescribe 2 grs. of
sulphate of atropia with 4 of acetate of morphia, to 1 oz. of
stiff glycerine ointment, and I tell the patient, to spread
about the size of a small filbert, on a bit of Mackintosh calico,
and to apply it to the epigastric region. This can be taken
off the following morning, and reapplied to the same region,
after spreading some more ointment on the calico. When
the atropia affects the eyes, the application should be omitted
until vision is fully re-established, when, if necessary, it can
be reapplied, every second or third night.

Extract of belladonna in frictions, and taken internally in 1 gr. doses, has already been recommended by Dr.
A. Bayle in his *Eléments de Pathologie Générale*.

If, besides the sensation of prostration, there be downright
pain, resisting the local means previously detailed, I have
sometimes, with benefit, applied a piece of lint steeped in
chloroform, and covered with oil-silk, to the pit of the
stomach, retaining it in its place by a bandage. A blister,
though often ineffectual, has sometimes relieved the pain;
and this reminds me that Comparetti and Barras derived no
benefit from blisters in similar affections they attended, at
Venice and at Paris; neither did Lorry. In the worst cases
of cardialgia, where there were fits of agonizing epigastric
pain, its intensity was often abated by taking from 30 to 60
drops of aromatic spirits of ammonia in the smallest possible quantity of water; or from 10 to 15 minims of chloroform on a lump of white sugar, and melted in a little water;
or by chloric ether; or by 20 drops of sulphuric ether

in one ounce of which a drachm of camphor had been dissolved; or by a drop of essential oil of peppermint, on a lump of sugar, and dissolved in water. In some cases the paroxysm would take its course, and abate of itself; in others, all milder remedies failing, I have denuded the skin of the epigastric region to the extent of a crown piece, by strong blistering tincture, or placed on its surface a piece of linen wetted with strong spirits of ammonia, applying afterwards from 3 to 6 grs. of acetate of morphia, and I have also derived benefit from the hypodermic injection of morphia. Veratria and aconitine are also invaluable agents in difficult cases. I have employed them externally, as Dr. Turnbull has done in other affections, incorporating the active agent with lard, 1 scruple of veratria, or 1 gr. of aconitine to 1 oz. lard, directing a piece, the size of a filbert, to be rubbed on the epigastric region, for a quarter of an hour, until warm and pricking sensations were excited, and this may be repeated, every second or third day. Trousseau and Mathieu thought highly of this method of applying veratria; and a liniment composed of ext. belladonnæ, ʒss.; tinct. aconiti, ʒiv., and liniment. saponis comp., ʒiss. may be useful.

If there be nervous irritability and sleeplessness I give one or two 3-grain pills of ext. of hyoscyamus every night, and if that does not do, 5 to 10 grains of Dover's powder, or 20 minims of the British Pharmacopœia solution of acet. of morphia, with 20 minims of ipecac. wine, and in exceptional cases the 16th of a gr. of acet. of morphia every one or two hours, until the induction of drowsiness.

Dehaen, Barras, and many who have paid attention to the obscure affections under consideration, praise equally with myself, the internal and external exhibition of opium, hyoscyamus, camphor, and cherry-laurel water. Dumas, of Montpellier, frequently prescribed the 15th part of the following compound to be taken daily, for nervous affections of the abdominal organs :—

Castor . . . . 30 grs.
Camphor . . . . 15 „
Opium . . . . 8 „
Conserve of roses . . . . . . q.s.

Indian hemp alone, or associated in pills with extract of henbane, is invaluable to quiet nervous symptoms ; the same may be said of bromide of potassium, and I have given ten grains of chloral three times a day with benefit, and I think it will turn out to be a valuable ganglionic nervine. Neither should I omit to mention, while writing of sleeplessness, in connexion with gangliopathy, that a friend of Dr. Handfield Jones has found a sinapism to the epigastrium, to give sleep when many other remedies had failed to relieve the insomnia of insanity.

Emetics are heroic remedies in a host of nervous affections, in long-continued globus hystericus, in spasms of the œsophagus, in nervous aphonia, in hysterical pains, or semiparalysis of limbs ; and as they frequently cure, without the removal of bile, they probably act by suddenly breaking in upon a morbid condition of the cerebral and ganglionic nervous forces, and promote their healthy equilibrium. When, however, there is great debility in connexion with the distressing epigastric sensations which I have described as gangliopathy, emetics do harm, and Joseph Frank, who was in Vienna, when Stoll's theory was in full force, mentions having seen the ill effects of emetics in such cases.

With regard to tonics, when gangliopathy predominates, I rely on good food, wine, and mineral acids, while the menstrual flow is still making irregular appearances. Mineral acids improve digestion, restrain irregular hepatic action, and cool the skin, often teazed by heats and flushes, and in support of my practice, I may mention that Hufeland gave Haller's elixir in diseases of the change of life. Acetate of lead, in one-sixth of a grain doses, has been frequently given, with success, by J. Frank, in cases of cardialgia caused by onanism, and accompanied by spermatorrhœa; but I have

never tried it.   J. Frank considers that the principal remedy
for neuralgia of the cœliac plexus is oxide of bismuth, the
third of a grain being given two or three times a day ; and
that the " oxide of bismuth and nitrate of zinc have a decided
influence on the nerves of the stomach, appeasing the nervous
affections of the cœliac, and of the abdominal ganglionic
plexus." Hufeland writes favourably of hydrocyanate of
zinc, in 1 to 4 gr. doses, two or three times a day, but I prefer
the valerianate of zinc.   Nitrate of bismuth is frequently
given in 10 to 20 gr. doses, suspended in a little mucilage,
and should be taken on an empty stomach.

Oxide of silver has been given by Sir J. Eyre and Dr.
Davey ; in recommending nitrate of silver, Dr. H. Jones
correctly remarks, that, when only taken for a fortnight,
there is no danger of the skin being darkened. Dr. Shearman
gives half-a-grain of nitrate of silver, with the same quantity
of opium, and 5 grains of ext. of camomile, three times a
day ; sulphate of quinine, dissolved in an acid mixture ; and
ʒj. doses of solution of potash, to prevent the formation of
sugar in the urine.

There is a general concordance of belief on the part of
good observers in the power of strychnia to rouse and increase
nerve force, and I give it when great loss of nervous energy
is not associated with anæmia or chlorosis. The powder of
nux vomica in 2 gr. doses, or its ext. in a 1 gr. dose, was
strongly recommended by Schmidtman. Dr. Davey gives
strychnia, and Dr. H. Jones writes encouragingly of it.   I
prize arsenic more and more as a nerve-tonic, at the change
of life. Dr. H. Jones praises it in cases similar to those I
have been describing, and if Dr. Leared has found it so useful
in cases of confirmed dyspepsia, it may be the drug acts
by toning the nerves, in and around the stomach. I have
seldom given digitalis at the change of life, but its power to
strengthen and regulate the heart, when given in small doses,
fully warrants its more extensive trial in visceral neuralgia.

When, after cessation, there is a tendency to anæmia or
chlorosis, I give quinine and iron, varying the preparations

in the same case, but never giving large doses. Comparetti and Barras speak highly of ice, to calm pain in spasmodic affections of the stomach, to increase the sensation of strength, and the peristaltic movement of the intestines. This reminds me that, in some severe cases, when the patient could not swallow cold water, or anything warm, there was a strong inclination for ices; which, as well as Nesselrode pudding, could be easily swallowed, and the pain was for a time relieved. Intense heat is as serviceable in this, as in other neuralgic affections. The actual cautery and moxas to the pit of the stomach have been used in olden times, so transcurrent cauterization might now be useful. The Chinese relieve cardialgia, by plunging acupuncture needles into the epigastric region. And lastly, there is electricity, for the application of which we have Pulvermacher's chains, one end of which is to be applied to the pit of the stomach and the other to the corresponding point of the spinal column or to the nape of the neck. Any other mode of applying electricity to the epigastric region, cannot be done too cautiously, or the result might be fatal.

CHAPTER VII.

DISEASES OF THE BRAIN.

TABLE XXV.

Frequency of Cerebral Diseases in 500 Women at the Change of Life.

Nervous irritability	459
Headache	208
Monthly headache	7
Sick headache	92
Hemicrania	3
Apoplexy and hemiplegia	6
Pseudo-narcotism	277
Hysterical state, or hystericism	146
Hysterical fits, since cessation only	3
Globus hystericus, since cessation only . . .	17
Laughing and crying fits, since cessation only	4
Monthly hysterical symptoms	1
Hysterical flatulence	8
Prolonged fits of unconsciousness	3
Epilepsy, since cessation only	3
Epilepsy, much aggravated	5
Delirium	3
Insanity	16
	1261

Thus 500 women divided 1261 forms of cerebral disease, many of them presenting several of these nervous symptoms at the same time, and this confirms the general belief in the frequency of cerebral affections at the change of life. After cessation, the liability to nervous affections greatly decreases; headache, nervous irritability, and pseudo-narcotism, have seldom the same intensity; and if cerebral hæmorrhage, apoplexy, and softening of the brain become more frequent, it depends on pathological conditions common to both sexes, in old age. A glance at the preceding table will show that at the change of life, women suffer chiefly from functional diseases of the nervous system, which explains why the mortality from cerebral diseases, according to the Registrar-General's reports, is always greater among men, than women from the twentieth year and upwards.

Of those who confirm my assertion, Gardanne and Dr. Ashwell state, that even when most favourably passed, the change of life brings with it a train of nervous symptoms. Gardanne and B. de Boismont speak of the tendency to headache; Dr. G. Bedford alludes to the frequency, the intensity, and eccentricity of the nervous symptoms at this epoch; and B. de Boismont thinks transitory delirium and insanity then of frequent occurrence. I believe that many women, are thoroughly unhinged by the change of life being left to take its course. Eccentricity embitters their existence and the lives of those around them; and should they lose fortune or friends, they may become insane, or may be thought so, erroneously, by relatives; hence arise scandalous suits, and the squandering of fortunes among lawyers, while all this might have been prevented by the systematic treatment of nervous symptoms when they first occurred. I cannot repeat too often that this crisis, like puberty, weighs upon the subsequent lifetime of women, and, thus considered, the nervous symptoms of the change of life assume vast importance. I shall now pass in review—1stly. Nervous irritability, the groundwork of all cerebral diseases; 2ndly, Cerebral neuralgia; 3rdly, Pseudo-narcotism; 4thly, Hysteria; 5thly, Epilepsy; 6thly, Insanity; and to understand these affections better, I shall throw on them the light derived from their study, at other phases of the reproductive function, and, after giving facts and the results of experience, attempt to account for their production and discuss their treatment.

1stly. NERVOUSNESS AND MORBID IRRITABILITY.—By this is meant that the nervous system is more than usually susceptible to external impressions, such as cold, light, noise, to the stimulus of diseased viscera and to that of emotion. There are not only innumerable degrees, but various modes of nervousness. If patients are asked if they are nervous, they understand the term to mean hysterical, and are often indignant at the question, though they will readily own that they cannot bear the slightest noise, although not over-sensitive to emotional stimuli. In some, the temper remains unruffled, but they are ready "to jump out of their skin" at the jarring of a door, and are in agony at hearing the leaves of a book turned over. To some it is torture to hear others converse; others say, that on the most trivial occurrence, they "feel pulses beat all over them;" they "feel all of a tremble, all of a shake;" "my flesh feels so heavy upon me," said one Dispensary patient, and "sensation is my calamity, not pain," was the eloquent expression of another; while one could not "stop in church because she felt the people too close." The fidgets—the *anxietas tibiarum* of nosologists—are a very frequent expression of nervousness. They indicate *nervous plethora*, seeking to be relieved by exercise or muscular action. Although one cannot determine the anatomical conditions of the nerves, when they are the seat of nervous congestion, I think it should be admitted as much as the congestion of blood-vessels by blood. The nervous fluid is too subtle to be seen or felt, but from the effects produced, the mind can deduce the fact of its being too abundant. I first allude to this nervous irritability because it is the "materia prima" of all nervous affections, the basis on which they rest, and the soil in which they grow; but to speak of nervousness, as *hysteria* is to perpetuate the state of confusion out of which the pathology of the

nervous system has scarcely emerged. The last table shows
the frequency of nervous irritability at the change of life,
and I believe it to be then, as at puberty, of almost constant
occurrence, just as frogs are more susceptible to electrical
influence, every returning spring.

2ndly. HEADACHE.—Our poverty of language for cerebral
diseases, is shown by the fact that we are obliged to give the
same name, to an indisposition that does not interfere with
habitual pursuits, and to something that will keep a patient
in sleepless agony for two or three days. Headache is a very
frequent symptom at the change of life, as at puberty. It
occurred in 208 of my cases. Some like E. P., aged fifty-
five, never had a headache until cessation, but have ever since
suffered from it more or less. In another patient headache
was the prominent symptom at menstruation, parturition,
and at cessation; a dull heavy kind of pain, with pseudo-
narcotism which lasted more or less for six months. With
regard to its seat, it may be met with in the following order
of frequency—viz., in the temple and forehead, at the top
of the head, and at the occiput. The last-named places have
been noticed by Friend, Etmuller, and others, as most
habitual. I have rarely met with it in the occiput, though
if Gall's localization were correct, that should be its most
frequent seat. With regard to the nature of this pain it is
described as "shooting," "throbbing," "gnawing," "burning," or as "if the head were in a vice." It varies in
intensity, from that slight amount which merely incon-
veniences, to that agonizing pain sufficient to prostrate a
hard-working woman and make her lie by, for some days.
In ninety-two cases, the patients suffered from sick headache
without any signs of gastric disease, and mucus was alone
brought up; and need one be surprised at sickness sometimes
accompanying cessation, when it often accompanies puberty,
and sometimes attends each menstrual flow ? Headache may,
doubtless, be caused by plethora, but generally it is a nervous symptom and behaves as such. I have had frequent
opportunities of observing the alternation of violent hemi-

crania and gangliopathy, the one arising as the other abated;
I have also witnessed a marked antagonism between the
cerebral and the pelvic extremities of the spinal cord, the
violent headache subsiding as pain became intense in the
sacrum; and as a general rule those who suffer much from
cerebral, are less troubled by pelvic pains.

CASE 22.—*Periodical headache at cessation.*—Sarah T., a
tall, stout woman, with the capillary vessels of the cheeks
very apparent, looks nervous and as if going to cry, was fifty-
six in September, 1855, when she came to the Farringdon
Dispensary. The menstrual flow appeared at sixteen, and
left gradually at fifty-one, after dodging her for a year. She
then suffered for a twelvemonth from flushes and perspirations, particularly at meals; otherwise she was pretty well
all day, but at two o'clock in the morning a severe headache
generally awoke her from a sound sleep. This occurred " off
and on, for two years after cessation," says the patient, and
her account is confirmed by her daughter. The sudden
loss of her husband induced great debility, and caused the
flushes and drenching perspirations to return, four years after
their subsidence.

Dr. Teissier, of Paris, has published—*Gaz. Méd. de Paris,*
1851—a case of periodical hemiplegia. A lady, sixty years
of age, becomes unconscious every month since the cessation
of menstruation and at the period she had been accustomed
to menstruate. On recovering her senses one-half of her
body is paralysed, and her speech is affected. This continues
for several days, and gradually disappears, to return at the
next monthly period. Being naturally of a calm and tranquil
disposition, those about her know when the attacks are
coming on, by her agitation and restlessness.

CASE 23.—*Cerebral Neuralgia.*—I have related in my handbook of " Uterine Therapeutics," how, after having enjoyed
tolerable health till the fortieth year of her age, a lady began
to suffer from chronic inflammation of the womb, which
increased nervous excitement and occasionally caused hysterical fits and vomiting that resisted all approved remedies,

M 2

and only yielded to the application of an issue at the pit of the stomach.

On menstruation ceasing in her forty-eighth year, the patient still suffered occasionally from vaginitis and ex-ulceration of the cervix with a yellow discharge, but she became less nervous and had no more hysterical attacks, which were replaced by maddening attacks of pain in the head, coming on every second or third week, and obliging her to remain for twenty-four hours in a dark room. During three years a host of remedies were tried in vain; among those given internally were the salts of zinc, Indian hemp, and bromide of potassium, while chloroform, the deuto-sulphide, and the tetrachloride of carbon were tried as outward applications to the nape of the neck, and to the temples; an occasional application of leeches to the temples doing more good than any other remedy, while the neu-ralgic nature of the complaint was shown by the fact that no severe attack occurred during a six weeks' stay at Birken-head; the attacks returning when she came back to her own healthy and comfortable home. After this state of things had lasted about two years, Dr. Russell Reynolds met me, and we agreed to consider the case as one of those unsatis-factory ones, in which pain seemed to depend on perverted cerebral nutrition, the result of chronic disease and its un-healthy necessities, and he suggested alteratives and tonics, and whatever exercise could be taken. The sudden loss of a husband, leaving her less well provided for, than she had a right to expect, and subsequent pecuniary losses, will account for the cerebral attacks becoming severer and more frequent. They often lasted forty-eight hours, with little abatement of pain, with a considerable amount of salivation, and incessant desire to pass a very large amount of scarcely coloured urine; and as there were often two attacks a week, they told very unfavourably on the memory of the patient, and on her power of attention: and she had also lost flesh, although digestion had improved. Such was the state of the case when Dr. Livingston, of New Brunswick, wrote to suggest the trial of tincture of aconite in a solution of hydrochlorate of ammonia, and I first tried scruple doses of this salt, three times a day, for a fortnight without any good effect, and on the 30th of last October I tried the tincture of aconite, beginning with five drops, but pushing on rapidly to twenty drops three times a day. After she began to feel fulness in the throat, and tingling in the tongue, the attacks ceased for a month, when came a desperate one. The aconite was continued, and again there was no attack for another month, when two occurred. I increased the aconite to seventy-five drops in the twenty-four hours, and there were no attacks for three weeks, and then two in succession. I diminished the dose of tincture of aconite to twenty drops three times a day, as the larger dose was disagreeable to her. Although the aconite has thus checked the frequent recurrence of the terrific hard attacks, never a day passes without headache, and about three or four o'clock A.M. the patient begins to bathe her head with Raspail's lotion, which she has found to be the most effectual of all milder measures, and in the afternoon she feels better and brighter. The respite given to the patient, by the aconite, has permitted her to follow out more fully the advice repeatedly given to her, to get out every day, and to mix more with society.

A fortnight after a previous attack, came one which lasted four days. The skin was hot, the lips parched, the pulse at eighty, the eyes deep sunk. There was a sensation as of blood rushing to the brain, and this was but slightly relieved by six leeches to the temples, although the loss of blood was considerable. At the next attack she took a whiff of nitrite of amyl, which frightened her by the prostration it produced. Chloral in doses of twenty to forty grains gave a certain amount of unsound sleep, and a different kind of headache, with painful congestion of the sclerotics. Dr. Russell Reynolds advised placing the patient under the influence of atropia during the attacks, but the effects were so disagreeable that the patient would not long take it. He likewise advised the arseniate of iron, which was taken for two months; it did no good, and assuredly caused a spasmodic cough, which only desisted on lying down. Now this patient is taking a combination of sumbul and bromide of potassium, at Dr. John Ogle's suggestion, but with questionable utility. The right remedies are not always attainable, and doubtless a pleasant companion to enliven solitary meals, and plenty of travelling, would do more good than medicines.

3rdly. PSEUDO-NARCOTISM.—The following cases will show the meaning I wish to convey by this term.

CASE 24.—*Pseudo-narcotism.*—Mary H., very stout, middling stature, sanguine temperament, flushed face, heavy eyes, uncertain walk and attitude, with all the appearance of being intoxicated, was sixty, when she came to the Paddington Dispensary, September 24th, 1849. The menstrual flow appeared at eleven with giddiness and sleepiness; it continued regular and abundant, though always accompanied by "a drowsy giddiness," for which she was often bled and cupped. She married at twenty-one, and had a child at twenty-six. The dodging-time began at forty, and lasted until fifty-eight, when cessation occurred. During these eighteen years the menstrual flow was irregular, being sometimes very scanty; at others, a flooding. She was also troubled with leucorrhœa, had back pains, faintings, flushes, sweats, headaches, and was so giddy that she frequently tumbled down, and people thought her tipsy. When walking out, she was always obliged to get near the railings, so as to have something to hold by. She said she never drank anything but cold water, and her landlady confirmed her statement. At about fifty-eight, when the flow stopped, she was at St. George's Hospital for ascites and boils. Diuretics considerably reduced the ascites, and for the last six months she has suffered more from cerebral symptoms. Intense headache, with pain, and singing in the ears, pains in the eyes, chills, and flushes, are now most distressing; and the perspiration trickles down her face. Every twenty-eight days these symptoms are very troublesome, and remain so for a few days. She has been unable to do any regular work for five years. I had eight ounces of blood taken from the patient, ordered a brisk purgative, to be repeated every week, several glasses of cream of tartar lemonade to be taken daily, and warm pediluvia every night. By following this treatment for a few weeks, the case was greatly improved; but she would not allow the bleeding to be repeated, whereas if ever there was a case in which bleeding would have been beneficial it was this; for she was of full habit, bore bleeding well, and the menstrual flow had always been abundant; three to four ounces of blood should have been taken from the arm every month, for the first two months, then every third or fourth month, with baths, purgatives, alkalies, and mild sedatives. This is one of the most marked instances of pseudo-narcotism I have met with. It appeared with first menstruation, returned at every menstrual period, and completely incapacitated the patient from earning her livelihood for five years.

What name then is to be given to this nervous affection? If it be called congestion of the brain, there were no signs of this condition in the following case, in which the nervous symptoms were similar :—

CASE 25.—*Pseudo-narcotism.*—Mary G., thin, tall, with dark hair, delicate complexion, drowsy look, and, when roused, looks as if she expected to see something dreadful, was forty-nine and unmarried. The menstrual flow appeared at fourteen, and continued regular, with little pain or other disturbance until the last two years, when it became irregular, and like " brown cinder dust." She also became unusually nervous, " feels silly, as if her head were one dead lump," and " five minutes after putting things away, she forgets where she has put them." More than once, she has lost her way in streets well known to her. She sleeps all night, wakes unrefreshed, often falls asleep during the day; sometimes feels stunned, and loses her senses for an hour. She complains of a *hot* pain at the pit of the stomach, with frequent sensations of sinking, which pain is worse after eating. She will stop three days without food, and then eat ravenously. For the last six months the bowels have been costive, and for the last two years, she has been troubled with flushes and sweats. I ordered the comp. camph. mixt. before meals, the carb. of soda after, 3 grains of blue pill, and 2 of ext. of hyoscyamus every night, also 10 grains of Dover's powder, a large belladonna plaster to the pit of the stomach, with hot pediluvia every night. This was continued for a week, and I never saw greater improvement in so short a space of time. The eyes seemed at peace with surrounding objects ; she looked cool, comfortable, and self-possessed ; she had lost her strange feelings, and was less drowsy during the day. The improvement was lasting ; and I only saw the patient now and then, when there was a slight return of the head symptoms. Pseudo-narcotism is generally allied to other cerebral symptoms, but in this case it was not obscured by other symptoms; for there was no headache, sick-head-ache, or hysteria. Again I ask, what is the name of this nervous affection ? Congestion of the brain ? There was none. Anæmia? There was none. Neither can it be called hysteria, unless by adopting the common custom of calling *hysteria* any nervous affection that is little understood. I say that these patients suffered from pseudo-narcotism ; and if it be urged that it is useless to give a new name to ex-ceptional cases, I reply that this affection is common at all the critical epochs of woman's life, especially so at the change, when I observed it in 277 out of the 500 cases.

At puberty, girls previously lively and clever, often become stupid and useless when sent on a message, as they forget what they were sent for, or how to come back ; or when at home, they will let things fall out of their hands, and fall down in stooping to pick them up. Pseudo-narcotism is often very intense when the menstrual flow is either very painful, deficient, or completely absent. I have known intense pseudo-narcotism, resulting from amenorrhœa, mis-taken for an idiopathic affection of the head in a girl of twenty-one. Her head was shaven, and she was bled and salivated, to the ruin of her constitution. N. G., at menstrual periods, could almost sleep while walking, and once remained sixteen hours in a state of stupor, from which she awoke quite well. K. R., at menstrual periods, would remain for hours in what she called her "quiet fit," a state of self-absorption, unaccompanied by hysterical phenomena, or by convulsions. Such examples of pseudo-narcotism are fortunately rare, for they present the extreme of what may frequently be observed. The ordinary symptoms are, a great tendency to sleep, an uneasy sensation of weight in the head, a feeling as if a cloud or a cobweb required to be brushed from the brain, disinclination for any exertion, a diminution in the memory and in the powers of the mind. But whether slight or intense, the symptoms indicate different degrees of the same cerebral affection which I have described in the patient's own words. They represent the diminished degree of that hysterical coma, in which, says Pomme, some would have been buried alive had he not interposed. A high degree of pseudo-narcotism is sometimes the immediate consequence of connexion. This was always the case in a patient of mine during the first year of marriage. Sometimes she would recover from it, in half an hour or an hour, sometimes the morbid would merge into the natural sleep, being followed by headache and prostration of strength during the following day, and I have repeatedly met with patients in whom connexion brought on sound sleep which could not be obtained from drugs. During pregnancy, the milder forms of pseudo-narcotism are frequent. The heaviness of head, the dulness of intellect, the giddiness, the tendency to fall, which are often erroneously considered symptoms of plethora, used formerly to be treated by venesection. S. C. always knew herself to be pregnant by feeling heavy in the head, giddy, and by very sleepy sensations which augmented as she increased in size. In P. N. there was no pseudo-narcotism at the menstrual epochs, but much during pregnancy.

The milder forms of pseudo-narcotism are amongst the symptoms of over-lactation. Physiognomists will be struck by the appearance women frequently present at this period,

their uncertain step and tottering gait, their vacancy of
feature, the drowsy or drunken expression of the eyes, and
the efforts they make to recover their oppressed intellects
when aroused by a question. When asked what they feel,
they complain of " heaviness in the head," " a stupid
feeling," " a lump in the head," " a stupid headache,"
" a numbness in the head, not pain," of " feeling silly,"
" stunned," and " the head like a dead lump." Some talk
of " an impulse to fall forward," they " are often giddy even
in bed." Others complain of " senselessness," of " feeling
lost and bewildered," " of a temporary loss of wits," " of the
fear of going mad."

At cessation, the temporary loss of memory is sometimes
a most distressing symptom. Patients forget where they
have put their purse, keys, or things they are in the habit of
using, a few minutes after putting them away. " I write
everything down, or I should forget it," says one patient;
and another, " I am obliged to stop and ask myself if I
am doing what is right." One was so forgetful that she
habitually made her children remind her, some forget their
way home and even their own names. Patients say they
feel so " sleepy and dreamy-like," " heavy for sleep, but
without pain," " can sleep anywhere," " can hardly keep the
eyelids open," and are " so sleepy they could drop off their
chairs." In a few cases, however, I have noted *sleeplessness*,
as in M. C., who says, that her head feels so funny when the
time for the menstrual period comes round, that she cannot
sleep for three or four days. Some patients feel as if they
had taken too much, as if something had got into their
head; they are accused of being tipsy, when holding by the
rails they pass in the streets; they have been turned off as
drunk, or at least are not allowed to open the door to
visitors by a master anxious to save his servant's reputation.
Pseudo-narcotism is most frequent and severe at the change
of life, while hysteria is rare, but it may alternate with
hysteria or gangliopathy, as in M. J., who, at this epoch,
suffered severely from dull pains in the head, on the subsidence

of which she could sleep anywhere. The nervous system of some women is so sensitive to the disturbing influence of the ovaria, that they have begun to be pseudo-narcotized at eight, and this has continued, more or less, until cessation. In advanced life, pseudo-narcotism becomes less frequent, less intense; should the contrary occur, it denotes a liability to insanity, as in A. H., who ceased fifteen years ago, but who is so sleepy she seldom rises before 5 P.M.

I have hitherto kept the results of my own observation clear from what stands on record : but lest it should appear that I have been too much struck by exceptional cases, I shall now show that esteemed authorities have met with cases similar to mine. They have not, it is true, drawn attention to the minor degrees of pseudo-narcotism, being satisfied to consider them as symptoms of hysteria, but they have noted the coincidence of prolonged sleep or coma with amenorrhœa, pregnancy, lactation, and cessation. Dr. Sandras, of Paris, gives instances of its occurrence in both married and unmarried chlorotics. Dr. Villarty de Vitré has made known—*Union Médicale*, tom. v.—a curious case of lethargy, occurring regularly, every month, in a girl suffering from amenorrhœa, the attack lasting seventy-three hours ; the lethargic state disappearing when the menstrual flow was restored. Dr. Copland notices coma, as a symptom of severe hysteria, in plethoric subjects, and in pregnancy ; and he mentions its occurring from the suppression of the menstrual and of the lochial flow. Montfaucon states, that in some cases of hysteria the principal symptom was continued sleep, lasting for several days, and only interrupted to take food. Pomme, who had a large acquaintance with nervous disorders, makes the following remark :—" L'hystérie est quelquefois accompagnée par une sorte de sommeil profond, qui prive les malades de tous sentimens. Elles perdent quelquefois connaissance aussi subitement que dans l'apoplexie, ce qui en a imposé plus d'une fois à ceux qui négligent d'examiner alors l'état de la mâchoire, qui est en convulsion dans l'accès hystérique." Tissot, and those who have written on onanism, relate that some of those addicted to it, do not sleep at all, while others are almost always in a state of stupor. Mr. A. Hunter—*Annals of Medicine*, 1799 —Mr. Blake and Dr. Montgomery have cited cases where pregnancy was accompanied by very great drowsiness; and in Dr. Montgomery's case the patient's memory was a perfect void, during the whole time of her pregnancy. Dr. Reid relates the case of a woman who was always able to judge pretty correctly of the time of conception, by a peculiar sensation of drowsiness, attended by sickness, with which she was then affected. J. Frank has known pregnant women to sleep eighteen out of the twenty-four hours, and he remarks that some women can distinguish fruitful connexion by symptoms which seem to imply a narcotic influence in the spermatic fluid; but independently of conception, I have known connexion give sound sleep when opiates could not. Baudelocque mentions the fact of women about to be taken with fits of eclampsia being thought drunk by the unexperienced; and although this disease may appear without prodroma, still heats and flushes, giddiness and bewilderment, are often its precursory symptoms. Soporous affections have been noticed at the change of life by Chambon, and Tissot says, " I have seen one of the most reasonable and witty women I ever knew pass two years of her life, at the cessation of menstruation, in a constant dream of a calm and gay character like her own habitual disposition. She was at the same time so troubled with the fidgets, that she could only remain sitting for ten minutes at a time, and if she persisted in doing so longer her sufferings were intense. Her nights were often sleepless, and remaining in bed was painful to her."

It is well to bear in mind that nature is immutable, that observers record the same facts, although they differ in their mode of interpretation, and I cannot better conclude these citations than by a case, related by Pomme, of periodical pseudo-narcotism in a young lady.

CASE 26.—*Successive fits of coma at menstrual periods.—*

Mdlle. ——, aged eighteen, sanguine and nervous, was suddenly affected with a lethargic drowsiness, when the menstrual flow should have come on, which symptom was alleviated by bleeding. At the following monthly period the drowsiness came again, but with greater intensity, and bleeding was again resorted to, but it failed to relieve her, so was not tried at the third and following menstrual periods, when the same lethargic state occurred, instead of the menstrual discharge. The drowsiness was followed by hysterical delirium, she became violent, and refused all food for seventeen days. Pomme put her, by force, into a tepid bath, and after remaining in it twelve hours, she quieted down, and took food. For two months she stopped daily eight hours in the bath, cold applications being made to the head; the delirium then disappeared, and the menstrual flow came.

The fact of morbid sleep being so frequent a symptom of the disturbed performance of the reproductive functions, tempts me to say one word on sleep as a physiological phenomenon. As sleep is often produced by the morbid action of the ganglionic nervous system on the brain, it is fair to ask, whether this nervous system has nothing to do with the production of our daily sleep? I think that a good theory of sleep should take into consideration the influence of the ganglionic nervous system in its production; and without attaching too much importance to a singular case, it may be mentioned that, in 1824, at the great hospital at Vienna, Joseph Frank saw a man, who for many years, passed all his time in sleeping. When his body was opened, the only unusual appearance was a considerable hypertrophy of the semilunar ganglia of the cœliac plexus, and of the other divisions of the great sympathetic.

I said that the singular cerebral condition described by me has no name, but is confounded with hysteria, with which it has nothing in common, or with coma, a disease of rare occurrence, which may be the result of some other affections. I have, therefore, given it a name, because new words are justifiable, when they serve to specify the nature of things.

Sir H. Holland has observed that, the difficulty of getting a correct nomenclature for morbid sensations, applies particularly to the cerebral. "Nervous stupor," or "spontaneous narcotism," would be good terms, but as the symptoms are similar to the gentle or powerful effects of narcotic poisons on the brain, I have applied to this group of cerebral symptoms the term *Pseudo-narcotism,* thus graphically expressing the fact, without prejudging the question.

4thly. HYSTERIA.—Medical writers should be careful to explain their meaning when using the term hysteria. When Dubois d'Amiens, Vigaroux, Béclard, Gardanne, and B. de Boismont say, women are not subject to hysteria at the change of life, and F. Hoffman, Pujol, and Meissner say that they are, what do they mean? Do they mean that they are not subject to globus hystericus, and the minor symptoms of hysteria, or that they are not subject to hysterical fits? A glance at the table that heads this chapter will enable the reader to decide the question relating to the frequency of hysteria at cessation; for out of my 500 cases, 146 who had suffered more or less from globus hystericus and the minor symptoms of hysteria still continued to suffer from them at the change of life; and, for the first time after cessation, seventeen had globus hystericus, and four had repeated laughing and crying fits, whereas only three had hysterical fits. In several of these cases, globus hystericus was so severe, as to make the patient jump up in bed, for fear of being suffocated. Landouzy gives a table showing that, out of 351 cases of hysterical fits, twenty-five only occurred after the fortieth year, and that they are very rare before puberty. Out of 821 cases of hysteria observed by Georget, Briquet, Landouzy, and Beau, only twenty-one originated from the fortieth to the fiftieth year, while 259 began from fifteen to twenty, and 574 from ten to twenty-five years of age. With regard to the effects of the social position, hysterical symptoms are much less common among the poor, and they occur most frequently when the nervous system of woman is wrought up to an artificial state by luxurious living, by overworking the mental faculties, and still more by the over-development of emotion. With regard to the influence of temperament in producing hysterical phenomena, I have met with them most frequently in patients of a sanguine temperament, in whom the least disturbance of the periodical function has, through life, often brought on this disorder; whereas in women of a nervous temperament, hysteria often diminishes in proportion as the activity of the reproductive organs diminishes, ceasing entirely with the subsidence of their action. Landouzy has shown that, in the vast majority of cases, hysterical convulsions were ushered in by pain or strange sensations in the hypo-gastric and ovarian regions, pains which induce the sensation of suffocation felt at the pit of the stomach, and then the globus hystericus. Dr. Copland has known the convulsive attack to be preceded by leipothymia, and I have often been struck by the alternations of gangliopathy with globus hystericus or other minor hysterical symptoms; globus hystericus subsiding, for instance, when a paroxysm of car-dialgia appeared.

CASE 27.—*Hysteria.*—Lucy P., tall, thin, with brown hair, grey eyes, and a chlorotic complexion, was forty-nine when she came to the Farringdon Dispensary, in October, 1854. She had always enjoyed good health, and married at eighteen; the menstrual flow first appearing the day after her wedding. She had several children, the last at thirty-three, and from that time until she was forty-four, the flow continued regular. It then dodged her for three years. For six months it came regularly every fortnight, then every two, three, or five months. Several times, when she had been long without the menstrual flow, " a gush of blood would come from the mouth or ears, and she was obliged to go to bed." She was much troubled with faintness, flushes, sweats, abdominal pains, piles, which bled occasionally, pseudo-narcotism, and for the first time in her life, hysteria. She once had a convulsive fit, and very frequently crying fits, and the sensation of a lump in the throat. After suffering in this way for three years, the menstrual flow completely ceased two years ago, and her health grew worse. She has become weak, and her complexion turned from ruddy to sallow; she is laid up for weeks with bilious attacks; she brings up bile, and her bowels are either relaxed or confined. She is much troubled with headache, with pseudo-narcotism, with nervousness, and with the hysterical symptoms already detailed. This patient had always enjoyed good health until the dodging time began; she had never been well for the following five years, but she was restored to health in about six months. She first took 4 grs. of blue pill, and 2 of ext. of hyoscyamus, every other night; the comp. camph. mixture before meals, and a drachm of carbonate of soda in a mouthful of water after meals. After three weeks these medicines were discontinued, and 5 grs. of citrate of iron were given in an effervescing draught, after meals, twice a day, and 10 grs. of Dover's powder every other night. The bowels were kept open by an occasional dose of sulphur.

E. W., aged fifty, the sister of a very nervous man, and herself very nervous, has not menstruated for the last sixteen months, but has been troubled by bleeding piles. She awakes every morning, with pain in the lower part of the abdomen. This is relieved by pressure, and by lying on the stomach; but only vanishes on her throwing up a large quantity of wind, which often gives her the appearance of being five months pregnant. There are no signs of dyspepsia about this patient. Under the influence of mental annoyance, I have often seen a copious extrication of flatus occur in healthy persons of both sexes, the flatus finding vent in repeated eructations.

5thly. EPILEPSY.—When it appears in connexion with first menstruation, it will often wear itself out after a certain number of years, but it occasionally returns at the change of life. I have seen four such instances, and in another, the patient first suffered from epileptic fits during lactation; she had seven children, but the fits only came on during four lactations, and lately at the change of life. She could assign no cause for their coming on at some lactations and not at others; but she had kept the temperance pledge four years previous to the epileptic fits at cessation. Drs. Tyler Smith and B. de Boismont have seen several cases wherein hysterical and epileptoid attacks only came on at first menstruation, at the decline of life, and at each menstrual period, the nervous symptoms completely disappearing on the cessation of the menstrual flow. Dr. Radcliffe informs me, that one of his patients had repeated epileptic fits at puberty, which induced medical men to advise her remaining single. She, however, married, had children, and did not suffer again from the epileptic fits until the change of life. In one of my patients, epileptic fits became more frequent and severe after the ménopause, and in two other cases, epilepsy occurred without any other cause than the change of life; and Moreau notes this epoch, as the only cause of epilepsy, in nine out of 529 cases occurring in women. Besides the well-known anomalous sensations arising from various parts of the body, I have met with one patient in whom the attacks were brought on by a fixed and intense ovarian pain, the result of subacute ovaritis. Pressure on the ovarian region often brought on attacks, and I myself thus caused a severe on. Dr. Copland has often seen epileptic fits preceded by leipothymia, the pulse retaining its usual strength; and this confirms my belief that the ganglionic nervous system may have something to do with epilepsy, although it be a disease of the cerebro-spinal system.

CASE 28.—*Aphasia.*—Mary H., a tall, healthy-looking woman, with brown hair, and hazel eyes, a harness-maker's wife, was fifty when she consulted me. The menstrual flow appeared at ten, and came regularly with little disturbance until she married at twenty-one. She had four children, the last at thirty-nine, and was regular until forty-seven, when cessation took place, after menstruation had been irregular for four months. Some days after the last menstrual flow, and without any known cause, while she was hanging out clothes in the garden, she felt giddy and fell down, remaining insensible for ten minutes. The fit was

N

repeated once a week, then every fortnight, afterwards with an interval of six weeks, when she became free from them for six months, but felt more nervous than usual; and some time after, while in apparent good health, and talking to her husband, she was all at once deprived of speech. She was perfectly conscious, knew what she wished to say, but could not utter the words. She had no pain, no choking sensations, she did not stammer nor stutter, she had no headache, but felt giddy even when lying down. For two years this fit came every day, and if it missed the day, it would come at night, on waking after her first sleep. She consulted me because the fits have, of late, come two or three times a day. She is now very nervous, frightened at the dark, and at everything, and she complains of headache. After well opening the bowels with the comp. col. and calomel pills, I gave my comp. camph. mixt. before, and the carb. of soda after meals. This was continued from the 12th of December to January 26th, and the number of fits of speechlessness diminished from four to one a day, and were of shorter duration. In addition to other measures, I then ordered 8 grs. of ox. of zinc, and 2 grs. of ext. of hyoscyamus to be taken every night. This was continued until March 15th, when the patient was discharged, still complaining of being once a day utterly unable to give utterance to her thoughts, but the speechlessness is of short duration. She was in good circumstances, had a kind husband, seemed in fair health, and the change of life could alone account for her fits, which resemble the little fits of epilepsy.

CASE 29.—*Epilepsy and Impetigo.*—Mary F., a woman of the average size, with grey eyes, brown hair, and a florid complexion, came to the Paddington Dispensary, February 16th, 1849. She was then forty-eight. The menstrual flow came at ten, and continued regular, sometimes too abundant, but without other disturbance. She married at seventeen, had three children, the last at thirty-four; since then she has been left a widow, and the flow remained regular until the last two years, when it began to dodge her. With the exception of giddiness during the previous year, she had enjoyed uninterrupted good health up to this period, but now flooding, at irregular periods, with great abdominal pain, lays her up for three months at a time; there is no uterine disease, but she is troubled with obstinate constipation, faintness, flushes, and sweats. She suffers from pain at the top of the head, which is worse when she lies down; without sick headaches or hysterical symptoms, she always feels giddy, stupid, sleepless, heavy, and "never as she ought to feel." Moreover, two years ago, when the menstrual flow began to be irregular, she had, for the first time in her life, a cutaneous disease, attacking both arms and face. The arms soon got well, but the face was sometimes better and sometimes worse. Now the cheeks are hot, livid, hard, and covered with flaky scales, a quiescent state which has lasted for three months, but the skin often blisters and distils a watery or a gummy discharge every month or six weeks. Since this affection of the face has appeared, the patient has been subject to fits, which come on suddenly from fright, exertion, or without cause. Talking to me has often brought one on. They come three or four times a day at uncertain periods. Her teeth chatter, she becomes unconscious, and this, at first, lasted only for a few minutes, but now often for an hour, when she becomes violent, and tears everything she can lay hold of, which obliges her to have some one to take care of her. The fits are worse and more frequent when the eruption fades, and the brain always feels relieved when the eruption is worse. February 16th.—I ordered 2 comp. col. and calomel pills to be taken every second night, and 1 oz. of castor oil the following morning; a scruple of sulphur and borax twice a day; one drachm of carbonate of soda after each meal. The comp. camph. mixt., hot foot-baths every night, 10 grs. of Dover's powder every night, were afterwards given, and March 8th, five leeches were placed behind each ear, for the patient would not be persuaded to be bled. March 17th.—The leeches did no good, the patient being no better than when she first came. In addition to the usual treatment, as the face is worse, it is to be fomented with warm milk and water. April 5th.—Abdominal pains were much complained of, and required enemata, with 30 drops of Battley's solution twice a day ; without pain, however, or any red discharge, she would frequently pass "large clean clots of blood." A blister once benefited her, so I ordered another, which was useless. June 14th.—She has just re-covered from a flooding, which had not occurred for the previous eight months. This has relieved the abdominal pains, but the head is as bad as ever, the nervous fits as frequent, and the face discharges much gum-like matter every second day. I ordered a blister to be kept open on each arm, but being myself laid up about this time, I lost sight of her. The life of no woman ever presented a more marked contrast. Perfect health up to the change of life, after which constant illness. The organic changes set on foot by the change of life caused all the patient's sufferings. Bleeding, daily tepid baths, prolonged for three or four hours, would have been very serviceable, but could not be taken. I would not have attempted to cure the cutaneous eruption before an abundant drain had been permanently obtained from blisters or issues.

6thly. INSANITY.—Some authors assert that insanity is more common to women than to men : and explain this supposed frequency by the shock felt in the nervous system at puberty and at cessation. The first assertion is doubtful, and the second an exaggeration : for, on consulting the Registrar-General's Reports, it will be found that deaths from insanity are most frequent in women from twenty to forty, or while the reproductive organs are endowed with greatest activity ; results in accordance with the statistics of Haslam, Pinel, Esquirol, and Fœdéré. From forty to sixty, when men are most actively engaged, and hope fails as well as physical strength, many more men than women die insane. From sixty to eighty, when the sexes most resemble each other, insanity affects them equally. A still better way of ascer-taining the liability to insanity at the change of life, is to take the admissions of a large lunatic asylum. The following table, for which I am indebted to the late Sir Charles Hood, shows the ages of women admitted at Bethlem Hospital, from January, 1845, to December, 1853.

TABLE XXVI.

Relative frequency of Insanity at different periods.

Period of Life.	No. of Cases.	Period of Life.	No. of Cases.
Under 15 years.	9	40 to 45 years.	162
15 to 20 ,,	61	45 ,, 50 ,,	153
20 ,, 25 ,,	216	50 ,, 55 ,,	122
25 ,, 30 ,,	223	55 ,, 60 ,,	57
30 ,, 35 ,,	217	60 ,, 65 ,,	55
35 ,, 40 ,,	218	65 ,, 70 ,,	27

Hence it appears that, although the time at which the change of life generally appears is not most prone to insanity, still 437 out of 1320 became insane from forty to fifty-five, whilst after that age the number suddenly diminished. Dr. Davey, of Northwoods, informs me that about one in every six insane women have had a first attack between thirty-six and forty, and that the chances of insanity diminish from forty to fifty-five. The effects of the puerperal state and lactation in producing insanity explain why it is most frequent in women before the fortieth year; but the table heading this chapter shows that, out of 500 women, there were sixteen in whom the mental and moral faculties were severely compromised by the change of life. B. de Boismont received into his asylum in one year eight patients, in whom insanity could only be accounted for by this cause. Dr. Forbes Winslow assures me that he has frequently seen insanity brought on by this epoch; and Dr. Wood confirms this assertion. If headache, pseudo-narcotism, hysteria, and epilepsy be caused by the change of life, it would indeed be strange if it should not likewise produce insanity. If Esquirol be right in establishing that derangements of menstruation form one-sixth of the physical causes of insanity, and B. de Boismont in professing the same opinion, and in saying that the menstrual epochs are always " *un temps orageux*" even for insane women who regularly menstruate, it would indeed be singular if the change of life did not sometimes produce insanity. If it be caused by puberty—a fact already recorded by Hippocrates— it would be astonishing if it were not sometimes the result of the change of life. A favourable prognosis of climacteric insanity may be generally given as to ultimate recovery, but it is best to explain to the relatives that the disordered state of mind may last for two or three years, although I have known women recover in six months. In two cases there had been insanity at an earlier period of life, but no relapse at the change of life.

FORMS OF CLIMACTERIC INSANITY.—I shall pass in review— I. Delirium; II. Mania; III. Hypochondriasis, or Melancholia; and IV. Irrepressible impulses, and the perversion of moral instincts.

I. DELIRIUM.—B. de Boismont, who has seen four cases of delirium at puberty, has likewise met with transitory attacks of it at the change of life. Sometimes the delirium is general, or it may run exclusively on one subject; and it is a matter of reflection for the psychologist, that while, in virtuous women, ideas thus combined, without the guidance of volition, may take a lascivious turn, in depraved characters, if Parent Duchatelet is to be believed, delirium runs on ordinary matters, and never on erotic subjects. I have met with three cases of delirium at the change of life.

CASE 30.—*Delirium.*—Mary S., of the average size, but thin, with a sallow complexion, dark hair, and hazel eyes, was forty-seven. The menstrual flow appeared at thirteen, and it had always been particularly free from morbid symptoms. Though twice married, she never conceived. About fifteen months ago the menstrual flow became irregular, and she was much troubled with headache and abdominal pains. One night she went to bed as usual, and awoke delirious. She ran down the street in her night-gown, required three men to hold her, was taken to the Bristol Infirmary, and in three days, the menstrual flow having come on, her senses returned. Several abscesses appeared in both arm-pits, some broke, others were lanced, and, when better, she came to London, and applied at the Dispensary for relief from headache, nervousness, and lightness of head. There had been no menstrual flow for the last ten months. I gave her a scruple of Dover's powder every night, and the comp. camph. mixture, which cured the patient. The singularity of this case is, that no nervous symptoms presented themselves until the attack of delirium. The patient was in tolerable circumstances, had a kind husband, and nothing to trouble her, so I cannot attribute the delirium to anything but the change of life.

CASE 31.—*Delirium.*—Mary L., a tall, stout woman, with brown hair, light grey eyes, and a sanguine temperament, was thirty-six when she came to the Paddington Dispensary, October 31st, 1850. The menstrual flow appeared at twelve, and continued regular, abundant, and without much disturbance. She had none of the usual head-symptoms of menstruation, but, from her twentieth year, was subject to epileptic fits. She married at twenty-two, had three children, pregnancy always warding off the fits. The menstrual flow was regular until her thirty-second year, when, a few days after its last appearance, without any known cause, she became delirious, and was taken to the Marylebone Infirmary. In a few weeks she was well enough to leave, and, though her health was not restored, she never had a similar attack. She is very nervous, giddy, and bewildered : but for the last year has had no fits. Ever since cessation the breasts have been very painful, and the nipples frequently exude a milky-looking or glutinous fluid. Treatment similar to that employed in the preceding case relieved the patient. She was forty when I last saw her, and the menstrual flow had not returned. If the sudden stoppage produced delirium in this case without any appreciable cause, the nervous system was tainted with epilepsy; the patient's sister was also epileptic.

II. MANIA.—Drs. Dusourd and Tyler Smith have noticed mania at the change of life. Before giving a curious instance, I shall mention that B. de Boismont has seen a case of mild dementia transformed by this epoch into furious mania, which lasted long. On the contrary, when cessation occurs in maniacs, it generally causes the disease to subside into dementia, and a sudden calm follows a state of furious agitation. Ferrus, Dubuisson, B. de Boismont, and others have noticed this singular effect, which reminds me of a mode of curing mania, ascribed to the priests of Cybele. *" Qui ante castrationem maniaci erant, sanam aliquanto mentem ab illo recuperant."*

CASE 32.—*Mania.*—Alice B., a bilious-looking woman, with gipsy features, dark hair, grey eyes, tall and slender, was forty-four. The menstrual flow appeared at thirteen; was abundant every two or three weeks, with a good deal of headache, sick headache, and pseudo-narcotism. Married at nineteen, had several children, the last at thirty-six, and was regular until she had a violent flooding at forty-two, since which time there has been no menstrual flow. After this sudden cessation, the abdomen swelled, was very painful, and without serious disease she dwindled down to a skeleton. She was improving, when, about two months after the flooding, as she was sitting by the fire, she felt a sudden flush in the head, face, and arms; she could not speak, and became unconscious. She was very violent, scarcely slept, eat enormously, and wanted what she could not afford. After this had gone on for three months, one night she said in a collected manner to her husband, "I will go to bed." She did so; slept soundly; was much more rational when she awoke, and gradually improved, without any other medicine than an occasional purgative. She never had a return of mania, but it has left her nervous, light-headed, and she forgets where she puts things. She came to the Dispensary for an inguinal abscess, with flushes, and drenching perspirations, and returned for advice whenever the nervous symptoms became too troublesome, and they were always subdued by scruple doses of Dover's powder, alkalies, purgatives, and tepid baths. The particulars of this case were confirmed by the patient's husband. Both state that the attack of mania came without any known cause; I therefore consider it a result of the organic changes evidently determined, in a nervous woman, by the sudden cessation of a flow accustomed to be abundant every two or three weeks.

III. HYPOCHONDRIASIS AND MELANCHOLIA. — These two degrees of the same mental condition are often met with at the change of life, particularly hypochondriasis, which seems to be an exaggeration of some of the symptoms of pseudo-narcotism. I have already drawn attention to the haziness of intellect, and to the state of temporary self-absorption into which women so often fall; to their love of solitude, their distrust of friends, their exaggerated estimation of trifles; and what is this but a temporary state of hypochondriasis, susceptible of becoming permanent at the change of life? This is why Gardanne and Dubois d'Amiens say they have often observed hypochondriasis at cessation; and they correctly remark that it is accompanied by epigastric suffocation, sensations of strangulation and neuralgia. Chambon has likewise noticed the frequency of hypochondriasis at this period, and thinks the bilious are most liable to it. Sir H. Halford has drawn from nature the following picture:—
" She sits in an indolent posture, looks gloomy, hardly speaks at all, and we learn from her attendants that she lives under the impression that some fancied evil is about to befal her. She is suspicious, undecided in all her movements, and manifests symptoms which differ in degree only from melancholy mania."

Dr. Maudsley also observes that " When positive insanity breaks out, it usually has the form of profound melancholia, with vague delusions of an extreme character, as that the world is in flames, that it is turned upside down, that everything is changed, or that some very dreadful but undefined

calamity has happened or is about to happen. The countenance has the expression of a vague terror and apprehension.
In some cases short and transient paroxysms of excitement
break the melancholy gloom. These usually occur at the
menstrual periods, and may continue to do so for some time
after the function has ceased. It is not an unfavourable
form of insanity as regards probability of recovery under
suitable treatment." In a paper read before the members
of the Provincial Medical Association, Dr. Conolly mentions
having seen the melancholia of cessation last two years ; but
those must have been very exceptional cases, of which he
could affirm that they were the beginning of the incurable
decline of bodily, as well as of mental health. With regard
to the causes of this state : it may be induced by plethora of
the portal system, by ovarian misrule, and uterine irritation,
but these physical causes would never produce melancholia
without some cerebral predisposition and the concurrence
of psychological causes. Peace and tranquillity of mind
may be the lot of those who have passed the crisis; but it is
easy to understand how the life of women, in this transition
period, may be replete with anguish. Supposing health not
undermined by the coming change, how can a sensitive mind
and a loving nature remain undisturbed when all is changing
around her, and one by one, the cords snap which anchored
her to life ? At fifty, parents may have been gathered to
the dust, children may have deserted the parental roof. The
flame of vitality cannot die without forebodings of decay, and
there springs up a doubt never before harboured—a doubt
whether, with faded charms and failing energy, one can
possibly retain possession of a husband's affection, and proofs
of unkindness are looked out for where none were meant.
Because the strength of youth is gone, a woman tries
to convince herself she is useless, and may become first
suspicious and then revengeful. If unmarried, is it wonderful
that this peremptory notice served by nature, to put aside all
long-entertained visions of fancied bliss should wound to
the quick a sensitive nature ? The future then becomes a void,

and despondency shows itself as boundless as the sands I have often watched at sunset, from my desert tent. In this desert of her thoughts no refreshing fountain is heard to pour forth the melodious song of hope ; no palm-tree promises relief against a scorching sun. She peoples the void with imaginary evils ; hears strange voices where all is silent ; feels awful forebodings, though nature smiles around her ; and thus hopeless and full of fear she will sit alone for weeks and months in the darkened room of some gloomy dwelling, without any other enjoyment than solitude, or that of brooding over unbegotten evils, with mental faculties now paralysed, or at times revived by conscience reproaching the poor sufferer with her inactivity, her sloth, and her want of faith in that God who deserts not His children.

CASE 33.—*Apathy and sudden change of previous habits.*— This incapability of, or rather dislike to, exertion,—this aversion to the mental exercise of *willing*, is characteristic of the female mind when disturbed by cessation. I have been often consulted by a lady blessed with connexions, personal appearance, and with a mind so highly cultivated that she might have taken the lead in society, had she not shrunk from its pleasures and duties, so soon as she first felt the influence of the change of life. Cessation took place five years ago, and ever since she has severed herself from connexions, shut her door to numerous friends, and lived in seclusion. She rises at 4 P.M. and goes to bed between 4 and 5 A.M., so that, in winter, she sees as little of the sunshine as the Laplander. She says she cannot do otherwise, though she knows that her mode of life forces her sons to seek elsewhere for society they cannot find at home. To expostulation or joking, she replies, "You do not understand me"; the usual reply of those who cannot justify their conduct. This want of energy, which has been so unfortunate for others, is no less detrimental to her own happiness ; for though annoyed at her distressingly nervous condition, she has not the courage to follow any plan of treatment. I said I had been often consulted, but my advice has been seldom followed, and few of the host of medical men this lady has consulted have been more fortunate. But what could medicine do in such a case without judicious management? Her present state depends upon her having had no strong guiding influence when it was most wanted. Evidently, therefore, to look for improvement while her actions have no other rule than caprice or apathy, is like placing an infant at the helm of a three-decker, and expecting it to steer safely into port. In this lady's case, a widow, with children unable to direct her, the greatest chance of recovery would be to enter the family of some judicious medical man, under a promise to stop three months, and implicitly to obey all directions. Thus would the patient, in spite of her lamentations and prophecies, be gradually brought back to reasonable hours; thus could the medical man teach her the long-lost art of taking exercise, sometimes mingling it with distractions, at others carrying it to fatigue; and travelling might complete the cure.

CASE 34.—*Melancholia, with suicidal tendencies.*—Mary W., a tall athletic woman, with a pale face, iron-grey hair, a whimpering tone of voice, and apparently always ready to cry, was forty-five when she came to the Farringdon Dispensary in November, 1855. She was the wife of a publican, and in good circumstances. The menstrual flow appeared at thirteen, and came regularly, even during lactation, for she had borne several children, but it ceased suddenly eight months before I saw her. Two months after cessation, she passed a large quantity of blood by the bowels, and for the last five months, every month or fortnight she has had several loose motions containing blood. The abdomen was also very painful and enlarged, so much so that she was thought pregnant by a high obstetric authority. For the last few months she suffered much from dry flushes during the day, and from " her skin stinging and perspiring" during the night. " All this," says the patient, "I could easily bear were it not for my nervous state." She complains of being all in a tremble, she is sleepless all night, and powerless all day, sometimes dozing, as if intoxicated, and waking up to thank

God she is still in her senses; at others, she sits alone, doleful and disconsolate, ashamed of herself for being so lazy, and still unable to do anything, or forgetful of what she ought to do. She says, that "when she sits thinking, she feels numbness and a pricking sensation in her limbs," and is much afflicted with suicidal thoughts. I first prescribed my usual mixture before meals, carbonate of soda after, 3 grs. of blue pill, with 2 of ext. of hyos. every other night, 10 of Dover's powder every night, a large belladonna plaster at the pit of the stomach, and vaginal injections with a solution of acetate of lead. February, 1856.—To relieve the abdominal pains, she took pills, each containing 2 grs. of blue pill, and a quarter of a gr. of ext. of opium. March 27th.—The patient was in every way better, but the motions were again bloody; the same treatment was continued, and on April 16th she stated that she was less subject to gloomy fits and to suicidal tendencies, though pseudo-narcotism was intense, the abdomen much distended by wind, and the tongue furred. An increased amount of flushes and perspirations coincided with the improvement, and she had grown suddenly stouter, although eating very little. I continued the same treatment, for I could not obtain her consent to have 3 oz. of blood taken from the arm. May 17th.—After fourteen months' absence of the menstrual flow, she passed, after great vaginal pruritus, many large black clots of blood, with skin-like substances, after which she had a watery discharge, and voided about a teacupful of blood by the bowels, with sickness, flatulence, and the old symptoms. More blood was passed on the 28th, after a scene with her drunken husband. I ordered her to take three times a day a tablespoonful of a mixture containing 2 drachms of the solution of hydrochlorate of morphia and of chloric ether, in 6 oz. of distilled water, and when this had cured the internal pains she took effervescing draughts three times a day. August 13. —She had again become bilious, had passed a little blood in motions, looked seven months pregnant, but her health and appearance have surprisingly improved, notwithstanding the worry of sick children, and a home ruined by the bankruptcy of an unkind husband.

CASE 35.—*Melancholia.*—Mrs. L., a thin, nervous-looking lady, with dark hair, and grey eyes, was fifty-nine when she consulted me in December, 1855. The menstrual flow appeared at fifteen, and came regularly with but little disturbance. She married at twenty, had six children, and was regular until the menstrual flow ceased gradually at fifty-four. Soon after this, without worry or any apparent cause, she became subject to headaches, drowsiness, and for a time both her sight and hearing failed her. She would sit by the fire all day absorbed in transacting strange things with strange people of former times, and often frightened by ghastly faces. At other times, she felt an incontrollable impulse to move about, and to dash her head against the wall, getting up and moving about without a motive, and then sinking down again, to remain immovable for hours. She slept with a light in the room, lest on awaking in darkness, she should throw herself out of the window. At times she had a strong impulse to kill her two grandchildren, and took care never to be left alone with them; but finding the temptation becoming too strong for her to resist, she left an affectionate daughter's comfortable home. She was much troubled, off and on, with dry flushes, and with the sensation of weight and gnawing at the pit of the stomach. I ordered 10 ounces of blood to be taken from the patient's arm, and, after a brisk purgative, I prescribed a quarter of a grain of acetate of morphia, every night for the first ten days, every other night for the next ten days, and every third night for the ten days following. I prevailed on her to stop two hours a day in a tepid bath, hot water being added when the water got cool. The baths were thus taken for a month, after which time, the patient looked and felt another person. All the symptoms had abated. I then gave 10 grains of Dover's powder every night, my comp. camp. mixture, and ordered a belladonna and an opium plaster to be applied, on alternate weeks, to the pit of the stomach, and a scruple of a powder composed of equal parts of flour of sulphur and of biborate of soda, to be taken twice a day. When this had been taken for a fortnight, the flushes, instead of being dry and burning, were mitigated by gentle perspirations. The treatment was continued for several months with slight interruptions, and a tour to the German spas completed the cure.

IV. UNCONTROLLABLE IMPULSES AND PERVERSION OF MORAL INSTINCTS.—As pseudo-narcotism has been seen gradually merging into hypochondriasis, so the symptoms of insanity are often those of hysteria, made intense and permanent. Girls, well-behaved until puberty, then become snappish, fretful, uncontrollably peevish, full of deceit and mischief: is not this a state of miniature insanity? Law has decreed, that to establish a defence on the grounds of insanity, it must be proved that, at the time of committing the offence, the accused did not know that he was doing wrong; whereas medicine teaches, that some are led by an irresistible impulse to deeds which they know perfectly well are criminal; deeds committed without motive, because liberty is warped by an instinctive impulse, which the patient cannot control. Freewill alone is punishable, says medicine; where freewill is not, punishment is barbarity, and the more barbarous, if inflicted by the hand of justice. It is however so difficult to distinguish the strong temptation, which might have been successfully resisted, from this irresistible impulse, that Judges are averse to the doctrine of irresistible impulses; they look upon it with suspicion, because it is the plea frequently brought forward to save undoubted criminals from the scaffold. Abstract views of punishment also render Judges averse to this doctrine. Baron Bramwell, in summing up on Dove's case, is reported to have said: "Why should punishment be administered at all? It is not inflicted on a man because he has inflicted evil on others, but to hold out an example to deter others from evil. That is the true reason, in my opinion, and the only object of punishment." Assuredly the punishment of criminals has also for its object the appeasing of divine justice, and the atoning to society for mischief done; and, although the inconsiderate admittal of the plea of irresistible impulse, would certainly increase crime by withdrawing that constant check on evil inclinations,—the fear of punishment,—still, rather than one innocent should suffer, it is better that a hundred criminals should escape. However much the plea of irresistible impulse may have been abused, I firmly believe it leads some to the commission of crimes they abhor. I have been consulted by at least ten women of high or low degree who, at the change of life, bitterly lamented their frequent temptation to commit suicide or murder. Judges, as enlightened as merciful, have admitted the doctrine of uncontrollable impulses in cases of puerperal insanity. If they admit that parturition determines uncontrollable impulses, they must also allow the possible occurrence of the same impulse at all the critical periods of woman's life; during puberty, pregnancy, lactation, the menstrual periods, and cessation. It is notorious, that the female mind is susceptible of being totally unhinged by the first impressions of the reproductive apparatus on the nervous system; by the irregular performance of the menstrual function; by connexion, parturition, and lactation; by some diseases of the ovaries and womb; and lastly, by the shock felt on the withdrawal of that ovarian stimulus to which the system had been accustomed for thirty-two years. Such being the case, what is to be the fate of women, who, at the change of life, unfortunately yield to some ungovernable impulse? Judges should be consistent, and as they shrink from inflicting punishment on mothers who murder their children during a fit of puerperal mania, and as on two occasions* they have advised juries to return a verdict of "not guilty" for the

---

* Regina *v.* Brixey, Central Criminal Court, June, 1845. The murder of an infant by this woman was proved; but she was acquitted on the plea of her being subject to disordered menstruation.

Amelia G. Snoswell was tried at Maidstone, March 20, 1855, for the same crime, and was acquitted on the same plea.

murder of infants by young women who had habitually suffered from disordered menstruation, they would, I trust, deal in a similar spirit with crime committed by women at the change of life, if it could be proved by medical testimony that this period has been an evident source of mental trouble, to a woman convicted of stealing, murdering, or committing other crimes. I lay some stress on this subject, because, when, a few years ago, a lady was brought up for stealing some pocket-handkerchiefs, and the mental infirmities determined by the change of life were pleaded in extenuation of the offence, the jury were divided. The difficulty of dealing with such cases is, that a patient cannot accept the benefits of a humane interpretation of the law, without abiding by the means wisely devised for the prevention of crime by the insane,—seclusion amongst criminal lunatics. Now seclusion may be useful in some cases, but is objectionable, until it has been proved that judicious treatment cannot remove a form of temporary insanity, as yet shallowly rooted in the system. Dr. Reid's* reflections on mental derangement apply forcibly to that occurring at the change of life. "Lord Chesterfield speaks in one of his humorous essays, of a lady whose reputation was not *lost* but only *mislaid*. In like manner, instead of saying a man has lost his senses, we should, in many instances, more correctly say, that they were mislaid. Derangement is not to be confounded with destruction ; we must not mistake a cloud for night, or fancy, because the sun of reason is obscured, that it will never again enliven or illuminate with its beams. There is ground to apprehend that fugitive folly is too often converted into a fixed and settled frenzy, a transient guest into an irremovable tenant of the mind, an occasional aberration of intellect into a confirmed habit of dereliction, by a premature adoption of measures, sometimes necessary, but only so in extreme cases."

I shall briefly notice the most frequent of the ungovernable

---

* Reid's "Essays on Insanity," p. 204.

impulses occurring at the change of life, after mentioning
that I have known women, previously economically inclined,
become reckless in their expenditure, and given to extrava-
gance, at the change of life; while others, from generous,
became penurious, or avaricious, in the midst of plenty,
and their talk was about the workhouse, and dying without
bread. A lady, who suffered much from melancholy at this
period, twice had *both* her ponies shot, because one of
them was taken ill; and a Dispensary patient laboured
under a delusion that she was covered with lice, which I
could never detect ; and, notwithstanding copious ablutions
and sulphureous baths, she constantly returned with the
same pitiful story.

TEMPER, OR UNCONTROLLABLE PEEVISHNESS.—As at puberty,
so at the change of life, temper is perhaps the most frequent
of its instinctive impulses. The patient often says, " From
mild and kindly disposed, I have become irascible and mischievous ;" or, " I have suddenly become so mischievous,
that I am quite afraid of myself." The character of a
charming lady, whom I have been so fortunate as to cure,
was thus, for a time, completely altered. The menstrual
flow came at thirteen, and left at forty-seven. During the
dodging-time, uterine prolapsus was troublesome, and required a pessary ; at cessation, there was uterine irritation,
incontinence of urine, distressing flushes and perspirations.
She was low-spirited, melancholy, snappish, quarrelsome ; so
uneven-tempered that her servants would not stop long with
her; and so suspicious, that a common-place observation
made by her husband, servants, or friends, was interpreted
as a conspiracy against her. This condition, after lasting
more or less for two years, yielded gradually to treatment,
and there has been a surprising amendment during the last
year. Both at puberty and at cessation this lady became suddenly very stout.

DIPSOMANIA.—Like B. de Boismont, I have several times
seen, in temperate women, a craving for spirits only at the
menstrual epochs, which subsided with the flow, and the

same desire has been noticed in pregnant and puerperal patients. Esquirol and H. Royer Collard have met with women, in good circumstances, who all through life had been temperate, but who, at the change, were suddenly seized with an irresistible desire for brandy, which again became disagreeable to them, when the critical period was passed. This impulse is akin to the well-known longings of pregnancy, and those who yield to it know they are doing wrong, struggle against it, but are overcome. It is easy to understand how such impulses should be rife at all the periods when the ganglionic nervous system is in a state of perturbation, and when anomalous sensations at the epigastric region indicate morbid action in the ganglionic centre ; they should, therefore, be considered not so much as despicable failings as complaints admitting of being cured by proper treatment.

KLEPTOMANIA.—Drs. Taylor and Marc have known patients who, previous to puberty or to disordered menstruation, were conscientious respecters of the rights of property, but who, though in affluence, would steal, at all risks, at the critical periods of life. Dr. Marc mentions a rich lady who, during pregnancy, could not resist the temptation of stealing a chicken from a cook-shop. I have already alluded to a case of this description occurring at the change of life, and I believe it to happen oftener than is supposed—though, while yielding to an ungovernable impulse, the sense of acting wrongly is still present to the mind.

HOMICIDAL MANIA.—This irrepressible impulse has been admitted in English Law Courts, whether it occurs during disordered menstruation, pregnancy, or puerperal mania. I cannot cite any case where this impulse led to lamentable consequences at the change of life, but one of my patients was constantly troubled with the temptation to kill her grandchildren, and she feared to dine with them " because of the knives."

SUICIDAL MANIA.—Hippocrates relates that self-murder was epidemical among the young women of Miletus. It has occurred during menstrual irregularities, and B. de Boismont has observed it at puberty and at the change of life. From his extensive statistical researches respecting suicide in France, it appears that, for one woman, three men commit suicide; and, with respect to the age at which this crime is most frequent, he found that, out of 5960 suicides committed by women in the whole of France, the greatest number, 1111, took place from forty to fifty; 1026 occurred from fifty to sixty, and 992 from twenty to thirty. It appears, however, that the capitals of some countries are exceptions to the rule, for the same observer found that, out of 1380 suicides committed by women in Paris, the largest number, 343, occurred from twenty to thirty, and 241 from forty to fifty. Capitals excepted, wherein the battle of life rages with fearful fury, it is safe to conclude that women feel the greatest propensity to self-murder between forty and fifty. Many patients have told me, with inexpressible anguish, that they feel "so strange in the head, so lost, so troubled with sensations of impending horror, that they must commit suicide to prevent their going mad," and one drowned herself in a cistern.

DEMONOMANIA.—When the belief in Satanic influence had a strong hold on the popular mind, lunatics often thought themselves possessed by the devil; now they are more afraid of the policeman. The only case of demonomania that I have seen, occurred to a lady at the change of life. She attributed the distressing symptoms of uterine disease to the devil having taken up his abode in her body, and the delusion vanished when her health was restored. Dr. J. Conolly relates a similar case, in his Croonian lectures; the patient was also at the turn of life: and on analysing Esquirol's remarkable article on demonomania, I am struck by the fact, that all his cases occurred at this epoch. One patient, aged forty-six, thought the devil had placed a cord from the pubis to the sternum; another, aged forty-nine, had been troubled by cerebral symptoms ever since cessation, at forty, and thought the devil lodged in her womb. A third, aged forty-eight, declared that he had taken up his abode in each hip-bone. A fourth, aged fifty-seven, from nervous had become insane at fifty-two, when cessation occurred, and she claimed the devil as the father of her children. A fifth, aged fifty-one, thought she had signed a contract with the devil—an illusion which originated in puerperal mania.

Impulse to deceive. — It is passing strange that women should surpass men by stupendous powers of deception. When man has an object to gain, when he wants to beg, or to escape military servitude, he is clever enough at deception; but he does not, like woman, find pleasure in deceiving for deception's sake. It would take a large volume to contain the authentic accounts of deception knowingly practised by women merely to excite interest. There was nothing at all amiss with the bodily health of most of these women, in some, menstruation was deranged, and others were hysterical. Thus admitting on the part of women a large amount of fully intended deception, there will remain a certain number of cases in which the patient is herself deceived, and has not the slightest intention to deceive others, even when making the most outrageous accusations. It is well known that irreproachable women, when suffering from puerperal insanity, have accused themselves of having had connexion with one or more men they knew little or nothing of. I have known women at the change of life to do the same, and who, after having been insane for six or eight months, have learnt with horror of what they had accused themselves.

APOPLEXY AND HEMIPLEGIA.—Notwithstanding Dusourd's contrary assertion, women are little subject to apoplexy at the change of life. I give the following curious instance of an ataxic state of circulation leading to the simultaneous loss of blood from many parts :—

CASE 36.—*Hemiplegia.*—Eliza C., aged forty-seven, first came to the Farringdon Dispensary, February 18, 1853, being then forty-four. She was of average stature and size, with brown hair, hazel eyes, and flushed face. The menstrual flow came at fourteen, and continued regular, but profuse,

and with unusually severe abdominal pains. She married at
fifteen. The menstrual flow ceased for eight months, and
then a large blood clot came away, and for several years
either the menstrual flow came every fortnight, or was absent
for five or six weeks, being then followed by a voluminous
clot and great flooding. She never conceived, was once
flooded for three weeks, and for the last six months the
menstrual flow has appeared every fortnight. Such being
her state of health, without known cause, she was suddenly
seized with hemiplegia of the right side. She walked to the
Dispensary, dragging the right leg, the right arm was numb,
often felt like pins and needles, and could be pinched without
causing pain. She looks stupid, complains of headache, loss
of memory, temporary loss of sight, stutters in speaking, and
cannot find the words she wants. The attack of hemiplegia
was accompanied by epistaxis from the right nostril. She had
been already cupped at the nape of the neck, and I ordered
8 oz. more blood to be withdrawn in the same way, and gave
calomel with opium, and black draught. February 21st she
lost blood from the right nostril ; 8 oz. blood were withdrawn,
and immediately afterwards came a violent flooding, which
continued until March 5th, when I ordered alum injections,
antimonials, and 1 gr. of opium, with 3 of c. ext. of colocynth,
to be taken at night. During the flooding, blood frequently
trickled down the right nostril. Notwithstanding the quantity
of blood lost by the patient, her strength seems but little impaired ; she is quite conscious, and walks to the Dispensary ;
but the hemiplegic symptoms are as marked as ever. May
12th.—She has returned from the country and is better, but
still complains of numbness of the right side. She can say
what she wants, but has a difficulty of utterance, and complains so much of pain on the right side of the head, that I
ordered her to rub in twice a day the size of a filbert of
2 oz. of mercurial ointment, mixed with 2 drachms of ext. of
belladonna, and to take the com. camph. mixture, and the
aloes and myrrh pills at night. June 16th.—After having
rubbed in the ointment four times, her head felt very sore,

and " as if it was heavier and larger than usual." She could
not see distinctly; salivation came on, but soon subsided,
after which she was better in every way, the head was no
longer painful, and she could use her right arm and leg,
though they remained weak. I lost sight of the patient
until May 31st, 1856, when she appeared the same as when
I first saw her, except that her face, instead of being flushed,
was spotted with dabs of capillary injection. Since the
flooding, during her previous illness, there had been no
menstrual flow, and she had suffered more or less from
headache, abdominal pains, faintness, flushes, and perspirations. Lately the head has been very painful, she finds it
difficult to keep awake, and cannot bear to be spoken to ;
the right arm and leg are almost useless ; the right hand is
soddened with perspiration, has pricking sensations, and
scarcely feels when it is pinched. Blood has been passed in
the urine, and she says that hæmaturia occurred also in her
first attack. I ordered 8 oz. of blood to be withdrawn as
before, calomel and opium to be taken at night, and a black
draught in the morning. June 25th.—She was cupped to
the same amount on the 4th and 18th, the bowels had been
kept freely open, antimonials had been administered, ice
applied to the head, and the patient was sometimes better,
sometimes worse, often bewildered, at times ungovernable,
but the hemiplegic symptoms remained stationary. I again
ordered the mercurial and belladonna ointment to the head ;
she rubbed in 2 oz., the head seemed swollen, she felt stupefied,
could not see clearly for three days, and was salivated. The
head and right side then became much better, but the right
hand is still powerless. She had a slight menstrual flow for
three days, after an interval of as many years, and I prescribed comp. col. pills, the camph. mixt., and 15 grs. of
nitre three times a day. Finding the hemiplegic symptoms
somewhat worse on the 16th, I advised the posterior half of
the head to be shaven, and the ointment to be again applied.
This was done with decidedly good effect ; and when I saw
the patient on August 20th, she no longer suffered from

hemiplegia, and was able to attend to her domestic duties, although suffering from debility, headache, and nervous symptoms, most likely increased by occasional fits of intemperance.

THEORY OF MENTAL DISEASE.—After recounting the facts which prove that there are morbid affections of the ganglionic centre, I sought to understand their import, before discussing their treatment; and now that the facts indicating the various forms of cerebral disturbance, induced by the change of life, have been related, I shall call attention to their probable mode of production, before stating the best means of cure. Those who only want facts, and eschew theory, may pass over what follows, until their eye meets the name of some familiar drug; but they should remember, that if practising in the time of Stoll, they would have inevitably sought to relieve the cerebral affections I have described by emetics; if during that of Mauriceau or Broussais, by repeated and copious bleedings. In other words, as practice must be swayed by theory, they had better choose the best.

How can organic or functional disease of the reproductive apparatus act on the brain? There are but two channels of communication between the reproductive apparatus and the brain; and something morbid must be conveyed from the reproductive apparatus to the brain either by the blood-vessels or by the nerves. It will not be difficult to show that the cerebral phenomena I have described do not depend on cerebral plethora, for, taking each phenomenon in succession, it will be found that intense nervousness exists more frequently in weak and anæmic than in plethoric patients. Leaving out of the question those in whom headache is caused by biliary derangement, it coincides much more frequently with a deficiency in the amount of blood than in the opposite condition, and often exists without any indication of congestion of the blood-vessels in the head. The same holds good with regard to hysteria and hysterical fits; doubtless there are sometimes fulness of the pulse and cerebral congestion, but in most cases the pulse continues weak, and the blood-vessels of the head exhibit no signs of over-distension. The symptoms described as pseudo-narcotism have many points of similitude with those of plethora and cerebral congestion, which were often, and are even now sometimes so interpreted, and when occurring in chlorosis or pregnancy, have been treated by bleeding; but the fact of the symptoms of pseudo-narcotism being often most apparent when the pulse is weakest, shows that it does not depend upon plethora. It is very seldom caused by biliary derangement, and its being unattended by paralysis, or other signs of cerebral disease, implies that it is not caused by structural lesions of the brain.

Doubtless, delirium and convulsions are often the immediate results of flooding; and when the cerebral symptoms I have described occur with chlorosis or profuse menstruation, they may be explained in the same way as the nervous accidents after flooding or too copious bleeding—the brain being very imperfectly stimulated by too small a quantity of watery blood. Very frequently, however, hysteria, pseudo-narcotism, &c., are observed when the tissues present every appearance of health, and when the amount of fluids in circulation seem in exact proportion to the wants of the system. The nervous phenomena cannot therefore be explained by anæmia. Drs. Todd and Cormack revived the old opinion, which attributes hysterical delirium and convulsions to a poisonous state of the blood, owing to the retention of something that ought to have been eliminated by the menstrual flow; but as hysterical symptoms are observed before first menstruation, they cannot be attributed to a poisoning of the blood. In this respect my observations are confirmed by Landouzy, who says, that he has observed symptoms indicating the influence of the generative organs upon the nervous system, long before first menstruation, and even before little girls had any idea of sex; and I have met with well-marked cases of pseudo-narcotism in girls of eight or nine years of age, though first menstruation was delayed to fourteen or fifteen.

If the reproductive apparatus does not act on the brain by

the instrumentality of the circulating organs and their contents, it must do so by means of the nerves. The genital
apparatus is richly endowed with ganglionic nerves, and I
have shown how frequently evident signs of disturbance in
the ganglionic centres coincided with headache, nervousness,
hysteria, and epilepsy. The influence of the generative
apparatus in the production of nervousness is distinctly per-
ceivable in many of the lower animals. In the beginning of
spring, just before the period of copulation, the nervous
system of frogs is endowed with a most remarkable degree of
irritability. The slightest touch will then excite those
states of the nervous system which, at other times, can only
be produced by narcotic poisons, or by strong galvanic action.
It is a matter of daily observation, that when women are
subject to increased ovarian action, they are also more irritable, more amenable to cold, to noise, to other physical
agents, and to emotional stimuli.

Catamenial headache may surely be considered a nervous
symptom; if it does not depend on plethora, is in frequent
connexion with other nervous symptoms, comes and goes like
those affections, alternates with them, and yields, like them,
to sedatives. Pseudo-narcotism may be explained in the
same way. It should be considered as a peculiar kind of
poisoning of the brain, by the too abundant galvanoid
influence sent to it, from the diseased ganglionic centre; or
else it may be owing to a loss of some control exercised in
health over the brain, by the ganglionic nervous system
similar to the acknowledged control of the same system over
the blood-vessels. Whether this disturbing influence acts on
the brain, or on some particular portion of it connected with
the function of sleep, I cannot say. Pathologists of all
ages who have particularly studied diseases of women, though
at variance in all else, agree in pointing to the reproductive
organs, as the starting-point of hysterical affections, and some
of them, with Landouzy, have observed that, in hysterical
fits, the patient generally complains, first of anomalous sensations at the lower part of the abdomen, then of pain or

suffocation at the pit of the stomach, afterwards of the sensation of strangulation, and lastly, of involuntary laughter or tears, convulsions, and coma.

The ovarian nisus seems to react on the *cerebrum abdominale*, so as to multiply its power, and causes it to so influence the brain, that woman, no longer the mistress of her own actions, is literally "fuddled with animal spirits, and made giddy with constitutional joy." The same ovarian nisus, acting with greater power on different parts of the brain, or on differently disposed nervous systems, after accumulating for a time, breaks out, spending its energy in hysterical convulsions, which may be followed by the temporary paralysis of the upper or lower limbs, as will be noticed in the next chapter. Finally, when the ovarian nisus is at the highest, if it be suddenly disturbed by intense mental emotion, the centrifugal nervous currents directing the menstrual flow receive a check. The whole energy of the ovarian nisus is thrown on the central ganglia, and reacts on the brain with such intensity, that in a few hours death ensues, and nothing is found but congestion of the cerebral blood-vessels—congestion, which may be in itself a result of the sudden shock. I would remark that, in ascribing such important results to nervous currents, I only follow the example of those who are working out the minute anatomy of the nervous system. For instance, Dr. Beale, in his paper on "The Paths of Nerve-currents in Nerve-cells," "concludes that the action of any nervous apparatus results from the varying intensities of continuous currents which are constantly passing along the nerves during life, rather than from the sudden interruption or completion of nerve-currents."

Pr. Schulzenberger, of Strasburg, has shown—*Gaz. Méd. de Paris*, 1846—that it is sometimes possible, by mere pressure on the ovarian region, to cause the radiation of pain from that focus to the epigastric region, and by continuing the pressure, to cause globus hystericus, and then disturbance of the brain and spinal marrow, hysterical convulsions being thus produced, while pressure on any other part of the body produced no such effects. In a highly nervous hospital patient, pressure to the ovarian region caused convulsions, without the intermediate minor symptoms of hysteria; and this experiment was repeatedly tried by several professors of the faculty of Strasburg, as well as by Pr. Schulzenberger. Similar cases have been seen by Romberg. The intimate connexion between the ganglionic and the cerebro-spinal symptoms of hysteria, their multitudinous gradations and great frequency, make me look upon hysteria as the keystone of mental pathology.

Esquirol has stated that the reproductive organs are often the centre from which emanates a stimulus sufficient to produce epilepsy; and I have lately seen this literally confirmed in a woman affected with subacute ovaritis, for pressure to the right ovary, intentionally or accidentally applied, repeatedly sent her off into epileptic fits, like the young lady mentioned by Sir B. Brodie, in whom a fit of chorea could be induced by the gentle pressure of the finger on the pit of the stomach. I have thus shown how the ganglionic nervous centre induces hysteria, which was already, to a certain extent, admitted by Willis, Van Swieten, and Lobstein. With regard to epilepsy, the predisposing condition of the cerebro-spinal system is, of course, different from that of hysteria, but the mode of induction of the two diseases is the same. Epilepsy may arise without the generative organs being at all implicated, some other viscus impelling the ganglionic nervous centre to diseased action. Little is known about the anomalous sensations and epileptical aura arising from various parts; but the epileptic fit is often preceded by intense nervousness and fretfulness, numbness or formication of the limbs, and sometimes by epigastric pain, or a sense of suffocation sufficient to show the influence of the *cerebrum abdominale* in the production of the convulsive fit. Some of the cases recorded in this work, particularly that of Ollivier d'Angers, show that a relation of cause and effect may sometimes exist between gangliopathy and paraplegia.

Having passed in review the effects of the reproductive

organs on the cerebro-spinal system at successive periods of life, it remains to be shown how, in some women, it can be the main cause of insanity. If the reader will recal the results of his own experience, he will own :—

1. That between the haziness of intellect, the slight forgetfulness of pseudo-narcotism and idiocy, there is no break ; that every intervening degree is exhibited in some women at one of the phases of healthy, or of morbid action of the reproductive organs.

2. That between the first slight estrangement of a girl's temper and the maniac's delirium there is no break ; every intervening link being supplied by some women at one of the successive phases of healthy, or of morbid action of the reproductive organs.

3. That between those first indications of spontaneous muscular action called " the fidgets," and the strongest convulsions of hysteria, there is no break ; every intervening link being supplied by some women at one of the phases of healthy, or of morbid action of the reproductive organs.

These organs can only react on the brain by means of their ganglionic nerves. What wonder, then, if the same powerful influence of the ganglionic nervous system should at times produce a *permanent* derangement of the mental and moral faculties, and permanent craving after what is sophistic in a mental point of view, after what is wrong in morals, as well as after brandy and physical stimulants? I am thus led to look on the ganglionic nervous centre as a source of vital power in constant correspondence with the brain for the maintenance of the " animal forces," as they are called, producing reflex variable morbid phenomena, in accordance with variable cerebral predispositions, when it loses its control over the brain, or otherwise exerts undue influence over it. If this be a true explanation of those rare instances of insanity produced by the undue action of the ovarian nisus, it follows that in other cases of insanity, its cause should often be looked for in the ganglionic nervous system, as well as in the brain ; for if one visceral apparatus, endowed with its gan-

glion and plexus can so react on the brain, through the inter-
medium of the ganglionic centre, as to produce mental derangement, why should not another, endowed with similar
ganglion and plexus, react on the ganglionic centre, so as to
cause similar results? In health, the ganglionic nerves of
each viscus send up few intimations of their operations to the
brain, and do not disturb the harmony of its functions; but
Hippocrates, and many other illustrious men, thought the
liver and connected viscera were the main cause of insanity.
Why should it not be so, since the converse holds good, and
since the mind, through the medium of the emotional powers,
which are twin-born with the intellectual, powerfully influences all the viscera endowed with ganglionic nerves? The
soul acts on the viscera by emotion, and they react on the
soul, so as to determine passion. There is no passion without
visceral—that is, ganglionic sensations. "The yearning of
the bowels," an oft-recurring expression in the Bible, is
physiologically true. Mothers feel it, so do lovers, and so do
sailors at the first sight of land. Neither can revenge be potent
without strong visceral reaction; or if, in popular phraseology,
"a man lacks gall." Broussais was right in saying, that
there cannot be epigastric pain without its causing some
shade of anger, which was partly determined by ganglionic
reaction. Many are without passion, because visceral sensations are slight, or absent; but if the brain acts so strongly
on the viscera, it is not surprising that they should, in their
turn, react upon the brain. When one of these viscera
becomes a prey to morbid action, it reacts by its ganglionic
plexus on the semilunar ganglia which influence the brain.
When the disturbance is slight, it is felt as a loss of power,
or what is termed "low spirits," or a sudden failure of mental
energy on feeling a sinking and faintness at the pit of the
stomach. From some slight visceral disturbance, lowness of
spirits and causeless melancholy frequently come over us like
a cloud. And if the cloud does not pass away, what is this
but hypochondriasis, or insanity, for which the cause will be
sought in the brain by those who only take a partial view of

pathology? These assertions bear the test of practice, and hold good with both sexes; for they are confirmed by some remarkable cases of insanity in men, some of which will be found in my work on " Uterine and Ovarian Inflammation." The patients were in good health, when, after a sudden mental shock, there immediately appeared the epigastric pain and phenomena, which lasted a considerable time, and were followed by a permanent cerebral disturbance, and by a strong and motiveless impulse to suicide and murder. Here the shock can be traced to the ganglionic centre, as in the following singular case :—

CASE 37.—*Melancholia and Gangliopathy.*—About 1860, a lady came from abroad to consult me. She was tall and thin, with kind and intelligent pale face, and was forty-eight years of age. Born in India, and brought up at Gibraltar, menstruation first appeared at ten, and was quite regular, and her health was perfect till she married at nineteen. Although there was no intromission, and the semen merely passed over her, and although the wedding-night brought on the menstrual flow, she *then* conceived; for there was a quarrel the following morning, and husband and wife parted never to meet again. This unexpected shock was immediately followed by tremulous sensations at the pit of the stomach, compared to the fluttering of a bird, and these sensations have continued ever since with variable intensity, being always aggravated by worry and bodily illness. When her son grew up, he went into the army, contrary to his mother's wish, and for the four following years there were superadded frequent fits of violent epigastric pain, with expuition of ropy mucus. After having been irregular for a year, menstruation ceased at forty-six, and all her sufferings had been worse for two years, and during that time constipation had been habitual, lasting once for sixteen days. The skin had always been dry and the flushes without any subsequent moisture. I found no signs of gastro-intestinal disease, no epigastric tumour nor aortic pulsation, and no uterine disease whatever. The fault was in the nervous system; for she had little sleep, with bad dreams, and an habitual distressing amount of nervous excitement or depression. She has vague terrors, is suspicious of everything and everybody, and is afraid of going to Woolwich, to see her son, and at other times she longs " to faint away, to be oblivious, to die, but not to commit suicide,"—although her father did so. She repeated, that " her life was a living death." The only treatment had been a three weeks' stay at Plombières, during the previous summer, and she had taken a daily warm bath of two hours' duration, and had drunk half a tumbler of mineral water twice a day, which had increased constipation. My advice was to keep the bowels open by purgatives; to take nitro-muriatic acid first, and then quinine, and a grain of Indian hemp every night for a fortnight, leaving it off for a week, then taking it again. I also wished plasters of opium and belladonna, to be alternately kept for a week on the pit of the stomach. If this plan of treatment did no good, I urged the patient to try hydropathic appliances, or the mineral waters of Aix-la-Chapelle, or Aix in Savoy.

There is in general no means of tracing the shock to the ganglionic nervous system, as in the previous case. The shock is spontaneous, self-generated, and seems to depend on some disturbance of the healthy action that one viscus exercises over the others, and all viscera over the brain, and I believe that insanity is generally preceded by more or less prolonged gangliopathy, but this is scarcely to be ascertained in lunatic asylums; to study insanity, in asylums only, would be like studying tubercular consumption in its second stage; the first stage of insanity is hidden in the domestic circle, incapable of understanding its phenomena, or anxious to hide whatever may be understood. This has been pointed out by Moreau, of Tours, who observes, " That almost all mental diseases are foretold and preceded by symptoms which generally pass unobserved, such as fainting, giddiness, and vertigo, and then by nervous sensations arising from different parts of the body like the aura epileptica, which the patients compare to excitement, or to electrical shocks." Dr.

Shearman has laid great stress on the frequency of mental disturbance in the cases he has reported, and the subsidence of the ganglionic symptoms on the occurrence of the worst cerebral symptoms, has not escaped B. de Boismont, who remarks, "Without denying the part of the brain in hypochondriasis, it is evident that this affection has its starting-point in the ganglionic nervous system ; and I have frequently remarked that gastralgic affections, with great disturbance of the digestive functions, *alternate* with mental diseases, and that these gastralgic affections cease entirely when insanity becomes permanent."

These views have been lately enforced by Dr. Maudsley, in his valuable lectures. "It is worth while observing," says he, "that in other forms of insanity, when we look closely into the symptoms, there are not unfrequently complaints of strange, painful, and distressing sensations in some part of the body, which appear to have a relation to the mental derangement not unlike that which the epileptic aura has to the epileptic fit. Common enough is a distressing sensation about the epigastrium : it is not a definite pain, is not comparable strictly to a burning, or weight, or to any known sensation, but is an indescribable feeling of distress to which the mental troubles are referred. It sometimes rises to a pitch of anguish, when it abolishes the power to think, destroys the feeling of identity, and causes such unspeakable suffering and despair that suicide is attempted or effected." And again he remarks, "Allowing that the generative organs have their specific effect upon the mind, the question occurs whether each of the internal organs has not also a special effect, giving rise to particular feelings with their sympathetic ideas. They are notably united in the closest sympathy, so that, although insensible to touch, they have a sensibility of their own, by virtue of which they agree in a consent of functions, and respond more or less to one another's sufferings ; and there can be no question that the brain, as the leading member of this physiological union, is sensible of, and affected by, the conditions of its

P

fellow-members. We have not the same opportunity of observing the specific effects of other organs that we have in the case of the generative organs; for while those come into functional action directly after birth, these come into action abruptly at a certain period, and thus exhibit their specific effects in a decided manner. It may well be, however, that the general uniformity among men in their passions and emotions is due to the specific sympathies of organs, just as the uniformity of their ideas of external nature is due to the uniform operation of the organs of sense."

Should I be asked to prove the preceding assertions by post-mortem examinations, I would reply, that at present morbid lesions are earnestly sought for in the brain, while the condition of the ganglionic system is never ascertained. Accurate research might confirm the assertion of Comparetti,[*] who found the splanchnic ganglia, particularly the semilunar, swollen and harder than usual in a hypochondriacal man who died at forty.

TREATMENT.—The frequency of slight cerebral affections, and the fearful gravity they sometimes assume, render their treatment most important. In the first place, it is the duty of the medical man to carefully ascertain whether the cerebral affection he is asked to treat is caused or aggravated by some visceral disease, for if a cerebral disorder is caused by uterine disease, or an affection of the liver, their cure is the first means that should be adopted to restore the healthy action of the brain; and if, by clinical facts and rigid argumentation, it has been shown that the cerebral phenomena of cessation do not depend on general or cerebral plethora, it is to be hoped that, whatever may be the oscillations of medical doctrines, the time will never return for headache, pseudo-narcotism, and some forms of hysteria to be treated

---

* Occursus medici de vaga ægritudine infirmitatis nervorum. Venetiis, 1780, p. 136.

by repeated and copious bleedings. As, however, the influence of plethora as an additional cause of cerebral affections at the ménopause is not to be underrated, I hope I shall not be taxed with inconsistency, if, while repudiating large bleedings, I seek to impress the conviction of the importance of occasional small emissions of blood at the change of life. Mild purgatives, either the saline, or vegetable extracts, with calomel or blue pill, are often as advantageous in cerebral affections, at cessation, as at other periods, even when they are not positively indicated by the furred tongue and morbid intestinal secretions. The use of antacids, cooling salts, and diuretics, as already explained, are useful adjuncts. Warm baths relieve the hyperæsthesia of the nervous system, severe pseudo-narcotism, and all those conditions which verge on insanity. The good effect of prolonged baths has been long shown by Pomme; and Recamier revived the practice, which is now generally followed in French lunatic asylums.

Hydropathic treatment will be found very useful in various forms of functional nervous derangement, particularly when the patients are stout and present no signs of organic disease. Witness a childless lady, the wife of a rich man, who, two years after cessation, persuaded herself that she was pregnant, on account of a slight increase of abdominal swelling. She would not allow an examination, for fear her anticipation should prove untrue; she engaged a monthly nurse, and made all other arrangements. After a time an examination was made, and the womb was found unimpregnated. For many months she was very hysterical, having choking fits, and long fits of crying and depression. Many remedies and doctors were tried without benefit, whereas six weeks of hydropathy restored her to good health, which has been maintained for the last three years.

Sedative medications are, however, the chief remedies; and what has been already said on this subject in the chapter on Therapeutics should be read as a preface to the suggestions about to be made. I will merely add that I have invariably

given sedatives for pseudo-narcotism, just as Hoffman cured
the sopor of intermittent fever with large doses of opium.
Faithful to the plan of applying the sedative to the suffering
surface, I seek to relieve the distressing headaches of the
change of life, and many of its more distressing affections,
by applications to the head.   In all mild cases of headache,
pseudo-narcotism, and hystericism, I recommend the patient
to sponge the head all over, once or twice a day, with cold
vinegar and water or eau de Cologne and water, and after
half drying the hair, to rub in, for five minutes, sweet oil, or
any pomatum.   If these refrigerant measures do not relieve,
hot flannels to the temporal regions can be tried.   The head
should be sponged with water, holding in solution 2 to 4
oz. of camphorated spirits of wine to the pint, with the
addition of a little eau de Cologne or lavender water.   Camphorated vinegar and water, or water in which camphor has
stood, or the comp. camphorated liniment well diluted, are
excellent remedies.   This last has been sold as "Ward's
essence for the headache;" but a still better preparation is
Raspail's sedative lotion, which is made by adding 2 oz.
of liquid ammonia and of common salt, and 3 drs. of camphorated spirits of wine to 32 oz. of water.   This lotion may
be used with a small sponge, or a pad of soft linen may be
soaked in it, applied to the painful part of the head, and
renewed as often as may be required.   It reddens the scalp,
causing burning sensations, but its action can be lessened by
diluting it with water, and in severe cerebral affections, a
handkerchief should be tied round the forehead to prevent
the liquid running into the eyes, while copious spongings
are made to the head of the reclining patient.   Cold cream
should afterwards be rubbed into the scalp, or cold cream
with 1 drachm to the oz. of camphor, and 10 drops of the
essential oil of bitter almonds.   In pseudo-narcotism amount-
ing to stupor, I have, in addition to other means, rubbed
into the scalp, eau de Cologne with as much camphor as it
would dissolve.   After rubbing it in for a few minutes the
patient has come to herself.   In a case in which these

attacks of stupor frequently followed the epigastric pain, this was my only treatment during the attack. On recovering her senses, the patient felt as if her brain were " benumbed," and then succeeded a sensation of internal pricking, like " pins and needles." When this was complained of, I wrapped the head in flannel, and left the patient to repose. The effects of these measures are sometimes surprising when they are faithfully carried out; but as nothing is so irksome as system, most patients prefer talking about the inefficacy of medicine to carrying out systematically any plan of treatment. In very exceptional cases the head should be shaven, and sedatives rubbed into the scalp. The ext. of belladonna, hyoscyamus, and opium, are what I formerly used, 1 drachm of each to 1 oz. of cold cream; and they may very advantageously be combined with mercurial ointment, as in the instances already given. The other treatment required to cure nervousness, pseudo-narcotism, hysteria, &c., is given in my chapter on Therapeutics and in the cases I have related. The case 23, at page 163, shows the many remedies that may be tried for cerebral neuralgia, and I may add that Graves found nitrate of silver to be useful in this complaint, and that it may be given for a fortnight without danger of darkening the skin. I will again remark on the urgency of procuring sleep, for long-continued sleeplessness is likely to produce insanity by habitually subjecting the mind to that increased intensity of feeling which takes place in the darkness, the silence, and the solitude of night. It is astonishing, in how much more lively a manner we are apt, in these circumstances, to be impressed by ideas that present themselves, than when the attention of the mind is dissipated, and its sensibility in a considerable degree absorbed by the action of light, sound, and that variety of objects which, during the day, operate upon our external senses.

In the first place, sleep should be courted by abstaining from exciting pursuits or amusements, between the last meal until bed-time. This will be often sufficient, and will allow re-

quired remedies to act well and speedily. It should be also
remembered, that if cold sometimes causes insomnia, this is
more frequently the result of the bad habit of so tightly
shutting up bedrooms that the air becomes foul and hot;
indeed, Dr. H. Bennet tells me that insomnia is to be always
cured by sleeping with the window open. Some cannot sleep
while the process of digestion is going on, while others sleep
better for a light supper, and very bad sleepers should
have a cup of milk or a slice of bread and butter, on their
night table, for taking one or the other will often give sleep to
those who have been tossing about for hours. Failing these
hygienic modes of procuring sleep, I have recourse to henbane
alone, or with Indian hemp, Dover's powders in 5 or 10 grain
doses, are the remedies I have most frequently prescribed ; but
chloral promises to eclipse all other agents when the only
object is to produce sleep, for its action is trustworthy and unattended by the drawbacks of all narcotics. It must not
be supposed, however, that opium and morphia are to
be dispensed with. They have other modes of action beside
that of promoting sleep, and are often useful at the change of
life. With regard to sleep, young practitioners must be aware
that the assertions of aged patients cannot be implicitly
relied on unless they be corroborated by other testimony;
for the aged often fancy they do not sleep at all, and
firmly deny having done so, after a very fair proportion of
that blessing.

When incipient insanity is accompanied by signs of ovario-
uterine disease, a very important part of the treatment is the
exhibition of sedatives by the rectum. Amongst others benefited by this mode of treatment, I may mention a patient sent
to me by Mr. F. Brown, of Chatham. She was about fifty
at the change of life, lived in complete seclusion, and was a
prey to all sorts of strange delusions. As there was leucorrhœa, frequent uterine pains, and great suffering on digital
examination, there was a chance that sedative enemata might
afford her relief; so I ordered them, with other measures,
though without giving much hope to her husband. I heard,

however, that the treatment had been very successful. Dr. Ferrus attributed great utility to menstrual medications, when insanity has coincided with, or has seemed to be determined by, the suspension of the menstrual flow; and he agreed with Dr. Conolly that, in such cases, the prognosis is most favourable. Many patients are sleepless, restless, nervous in the extreme, always in motion, always attempting, but with little strength to perform; the distinction, then, between the radical and acting forces should be remembered, the object being to moderate the acting, and to increase the radical forces; accordingly, steel may sometimes be advantageously associated with sedatives; the citrate of iron in 5 or 10 gr. doses, in an effervescing draught after meals, is the best way of taking it. With regard to the prevention of nervous affections and insanity at this period, my only recommendation is, to read over again the chapter on the general principles of hygiene at the change of life.

CHAPTER VIII.

TABLE XXVII.

Liability to Neuralgic Affections at the Change of Life.

Lumbo-dorsal neuralgia	226
Hypogastric pains	205
Monthly hypogastric pains	16
Frequent ovarian pains	5
Habitual dry colics	6
Paraplegia, mild forms of	6
Sciatica	4
Intercostal neuralgia	6
Numbness and impaired sensation in hands and arms	3
Paralysis of arms, hysterical	1
Brow ague, brought on	1
,, ,, increased	2
Aphonia, hysterical	3
Temporary deafness, induced or aggravated .	10
Permanent deafness	1
Odontalgia	2
Total	497

Thus, during the change of life, 497 neuralgic affections were divided amongst 500 women, some taking more than their share. In after life, these affections become much less frequent, with the exception of lumbo-dorsal neuralgia, which often persists, in a slight degree, until advanced age, and few women escape those neuralgic affections of the nervous expansions of the skin, which are often called rheumatism, while of course the senses become more and more obtuse. When the organs of vegetative life are seriously disturbed in their functions or structure, pain is experienced, sometimes in the viscera themselves, but most frequently in some portion of the walls of the cavities in which they are contained. The spinal nerves which are distributed through the viscera, receive the morbid influence, transmit it to that portion of the spinal column whence they originate, and the pain is then reflected through the spinal nerves which proceed to the cavities containing the viscera.

The ovary may transmit pain to the spinal nerves by means of the splanchnic nerves. The upper part of the womb is supplied with spinal nerves from the intercostal branches, through the medium of the splanchnic nerves and ovarian plexus, and any disease seated in that part of the womb may cause the reflected pains to be felt in various parts of the spinal column. The middle and lower portion of the uterus is furnished with branches of spinal nerves from the lumbar plexus through the medium of the hypogastric, and when this part of the womb is diseased, the pains are transmitted along these nerves and reflected on those which arise from the lumbar plexus, and therefore along the nerves supplying the muscles of the lumbar portion of the back, the walls of the abdomen, inside of the thighs, the front of the leg, and even sometimes to the instep. The spinal nerves distributed to the vaginal portion of the generative intestine arise from the sacral plexus; hence disease of the vagina causes pains to be reflected along the nerves which come from this plexus, and as this plexus furnishes nerves to the sacral region, to the peritonæum, the posterior part of the thighs, and the calves of the legs, pains may be experienced in all this course, and, in some rare cases, even in the soles of the feet. It will therefore be seen that it is not possible to ascribe the dorsal and the hypogastric pains, each to a distinct set of nerves.

DORSAL PAINS.—These are almost always fugitive in the

upper portions of the spinal cord, and principally settle in its
lower extremity, radiating to the small of the back, the loins,
thighs, and legs. The pain is generally described as an
aching or numbing pain, a gnawing, dragging, burning, or
grinding pain; a sensation as if the back were broken, or as
if it were opening and shutting—varieties of pain, like those
of neuralgia in other parts of the body; their intensity varies
from that of slight pain, which does not prevent moving
about, to that which, for a time, usurps the place of all other
sensations, confining women to their beds for a few days.
The frequency of these pains is as follows :—

Lumbo-dorsal pains already existed in . . 70 per cent.
They were augmented at cessation in . 46   ,,
      ,,   the same in 17   ,,
      ,,   less in 7   ,,
They did not exist in   30   ,,
                                          ———
                                          100

A great intensity of pain in the brain and spinal column
is seldom met with at the same time, for the two are in
general so counterbalanced, that when a great amount of
cerebral disturbance exists, the spinal symptoms have not a
similar intensity. During the prodroma of menstruation,
and during the "dodging time," the cerebral symptoms are
most intense; the spinal symptoms are, in general, more
common and annoying during the period of the full activity
of the generative function; and at cessation great is the intensity of both modes of suffering.

HYPOGASTRIC PAIN.—This pain is generally referred to the
ovarian, and to the uterine regions. It differs from the
symptom just described, in being a pressing, forcing, or
bearing-down pain. It seems to indicate a tenesmus of the
cervix uteri, to have an expulsive character, and to mark the
direction of those neural currents which direct the course
of blood towards the womb, and procure its expulsion from
that organ. Even when the menstrual flow has ceased, these

pains sometimes recur monthly, and determine the leucor-
rhœal discharge, or the diarrhœa, by which they are often
accompanied. Such pains had previously recurred during
menstruation in

	51 per cent.	
They were augmented at cessation in . 30	,,
,,　　the same in 12	,,
,,　　less in 9	,,
There were none in 49	,,
	———
	100

I have known women suffer habitually from colics for
eight and ten years after cessation, the intestines being the
seat of pain, which, however, did not interfere with their proper
functions. The following was a tedious case in which dorsal
and a fixed ovarian pain were the prominent symptoms :—

CASE 38.—*Lumbo-abdominal neuralgia.*—Patience K. was
tall, stout, with a broad face, thick masculine eyebrows, dull,
squinting grey eyes, brown hair, a sanguine temperament, and she was in her forty-eighth year. The menstrual flow appeared at fifteen, and continued regular
without disturbance. She was once laid up for a year
by some acute affection of the brain before she married
at twenty-four, but was fruitless. At forty-four she again
suffered much from pains in the head, and pain in the
right arm, which was benumbed and contracted for some
time. This was cured, but P. K. has been ailing ever since,
off and on, subject to a throbbing, heavy pain at the top of
the head, to nervousness, to trembling, to loss of memory, to
palpitation, to epigastric pain, sometimes only after taking
food, at others without a cause. This has been more troublesome during the past year, and for the last four months she
regurgitated sour or bitter stuff once a week. She looks
vacant, bewildered, and of late has had fainty feelings, and a
clammy skin. For the last two years there have been frequent pains in the ovarian regions, " as if from the plunging

in of a knife," and for the last year intense pains have arisen at the lower part of the back and sacrum; her other sufferings have increased, and have become more frequent. During this time the menstrual flow has also been irregular, much more abundant, and with scarcely more than ten or twelve days between the menstrual periods.

To allay the derangement of the biliary functions was the first indication; the next was to prevent the too frequent recurrence of the menstrual flow, by adding 6 drachms of tincture of cinchona to the comp. camph. mixture. Two tablespoonfuls to be taken three times a day. I also prescribed 20 drops of liquor potassæ in a little water, after meals, and at night 4 grs. of blue pill, with 2 of ext. of rhubarb, to be followed by ½ an ounce of castor oil in the morning. An opium plaster to be applied to the pit of the stomach. Feb. 12th.—Ten oz. of blood were taken from the arm. The previous remedies were continued, with the exception of the pills, which were now to be taken only on alternate nights. March 11th.—The menstrual flow has not appeared for six weeks. The food is better digested, but the tongue is still furred, and there is much pain under the left breast and in the left ovarian region. Leucorrhœa is abundant, the os uteri has the usual size, it is painful on pressure, which also increases the habitually felt hypogastric pains. There is no ovarian swelling. April 21st.—A blister to the left ovarian region only relieved the intense pain for two days. Mercurial and ext. of belladonna ointments rubbed in twice a day did no good, so I ordered 20 drops of Battley's solution and 1 drachm of tinct. of hyoscyamus mixed in a little warm milk to be injected once a day into the rectum. May 20th.—The sedative injections have given great relief. The menstrual flow came without much pain after an interval of three weeks. There is still intense pain in the back. On again examining the patient I can find no organic uterine disease. The pains are caused by the cessation of the uterine function in a woman whose nervous system has been shattered for the last five years. July 22nd.—The menstrual flow came after an interval of ten weeks. The patient is again very bilious, and is ordered the blue pill and previously named remedies; 4 oz. of blood are to be taken from the arm. November 1st.—The biliousness has gone, the sedative injections control the intense pains; the head symptoms are better. The menstrual flow has again appeared after an interval of thirteen weeks. When I lost sight of the patient she had been many months without menstrual flow, the pains much less severe, and her health good.

TREATMENT.—There is a distinct indication to use sedatives until the habitual pains are assuaged, and they may be continued for weeks without ill effects. Soothing liniments may be rubbed in night and morning; I mean camphorated liniment, with the addition of tinc. of opium, belladonna, hyoscyamus, separately or combined, as the case may suggest. Soft opium or belladonna plasters may be applied and renewed every four or five days; and, better still, my ready-made plasters, made by spreading atropia glycerine ointment on Mackintosh calico, as described in my "Handbook of Uterine Therapeutics." Linseed-meal poultices sprinkled with camphor, mustard poultices, and even blisters, may be useful. For abdominal pains the best application is a piece of piline large enough to cover two-thirds of the abdomen, to the upper corners of which tapes are to be sewn, so that the piline may be worn like an apron next the skin during the day. This alone often gives relief, but a teaspoonful of laudanum sprinkled over the cloth side of the piline makes the remedy more efficacious.

Paraplegia.—In infants, paraplegia generally depends on intestinal irritation; in adults, on vesical or uterine affections; and in old age, it appears as an idiopathic disease. It might have been supposed from the manner and frequency with which the lower part of the spinal cord is influenced by menstruation, miscarriages, parturition, and diseases of the sexual organs, that women would suffer more than men from diseases of the spinal cord, but out of 177 cases of paraplegia tabulated by Brown-

Séquard, only forty-nine occurred in women. Out of 114 cases of locomotor ataxy collected by Dr. Topinard, there were only thirty-three women; and in seventy cases of the same disease collected by Eisenmann, there were only twenty women. With regard to paraplegia at the change of life, it must be considered a very rare disease.

Gardanne mentions having seen paraplegia occur after the sudden cessation of the menstrual flow, but he gives no details. My six cases and a few others subsequently observed were of a mild nature, and all recovered. There were pricking sensations in the feet, numbness of the lower limbs, great pain in the dorsal region, and an inability to walk. Three complained of a difficulty in passing water, and the sensibility of the skin of the lower limbs being impaired. Similar cases have been met with by Dr. G. Bedford and B. de Boismont. Paraplegia occurred in two out of Dr. R. Leroy d'Etiolles' twelve cases, and on careful inquiry at the Salpêtrière, where there is a great number of paralytics, many of the paraplegics dated their complaints from the change of life. Most of them might have been cured in the early stage of the disease, when it depended upon congestion of the spinal cord; but, subsequently, atrophy of its lower portion prevented the possibility of cure. In many cases, there was no organic uterine disease to account for the paraplegia; but in one of Dr. R. Leroy d'Etiolles, there was considerable uterine swelling, and abundant leucorrhœa, and the patient could not walk without several attendants. Iodide of potassium cured the uterine hypertrophy, and the paraplegia disappeared. The following case is related by Ollivier d'Angers, who does not seize the import of the ganglionic symptoms which play so prominent a part in the case.

CASE 39.—*Frequent attacks of paraplegia, with gangliopathy.* —A lady, aged forty-nine, in whom the menstrual flow had been very irregular for a year, every now and then, without ascertainable cause, had the following train of symptoms, whether in bed or up, reposing or walking, eating or fasting. There was pain and constriction in the epigastric region, with or without sensations of suffocation, which sometimes awoke the patient suddenly. Then came flushing of the face and head, and afterwards burning sensations in the upper part of the spinal column, with pain and numbness in the arms, which became partially paralysed. This lasted for about an hour, when cold perspirations appeared; she then felt as if cold water were poured down her back, and her lower limbs became paralysed. This temporary paralysis has often occurred so suddenly, that the patient, taken unawares, has fallen down. In about three-quarters of an hour, these symptoms disappear, without leaving any trace; but they have recurred, from three to five times in one day, on several successive days, previous to the menstrual flow, and about ten days after, during which time constipation was great, urine more frequently passed, and with tenesmus. Leeches repeatedly applied opposite to the painful part of the spinal column, tepid baths and purgatives, greatly diminished the frequency and intensity of the attack, and the menstrual flow became more regular.

A blow to the pit of the stomach determines paralysis; and here, the womb, after its fashion, gave a shock to the nervous ganglia at the pit of the stomach, which reacted on the spinal cord, and caused paralysis of the limbs. In remarking on this case, Ollivier d'Angers insists on the necessity of applying the leeches to the spinal column, asserting that the same effect would not have been produced by applying them to the pudenda; but this position is disproved by the following case, which occurred in the practice of a medical friend :—

CASE 40.—*Paraplegia.*—A lady's-maid, aged forty-five, complained of violent pain in the loins, for which a mustard poultice was ordered; and, as the pain persisted, a blister was subsequently recommended to the lumbar region. This application was soon followed by paraplegia, and a neighbouring practitioner gave, as his opinion, that the application of the blister had determined the paralysis of the lower limbs. Although this assertion was contradicted by another

medical man, who was called in on account of the persistence
of the paraplegia, my friend received several letters from the
solicitor of the family, menacing him with an action; but he
set them at defiance, and he afterwards learnt that the patient
went home to her friends, and that a country practitioner,
more clear-sighted in this instance than the eminent men of
town, putting together the circumstances of the patient's age
and the previous irregularity of menstruation, applied leeches
to the womb. The result was a gradual diminution of the paraplegia, and she was soon able to walk with perfect ease.

TREATMENT.—In how many cases, where partial paralysis
was supposed to be caused by inflammation of the spinal
cord, have the backs of delicate patients been uselessly
tormented by blisters and moxas; whereas, the treatment
should be directed to whichever organ, womb or kidney, may
cause the paraplegia. I have seen admirable effects from the
use of the hot mineral waters of Aix in Savoy, and great
good is sometimes done there by parboiling the powerless
limbs, the patient being placed in a bath, in and out of which
hot water is continually flowing. Shower baths and douches
of hot spring water will be found useful; and shower baths
of water, alternately hot and cold. Such appliances, irre-
spective of sex, are valuable, not only in paraplegia, but in
all chronic, local paralytic affections, which seem to depend
on some obstructed circulation of the nervous fluid. When
the patient cannot leave home, benefit may be derived from
stimulating baths, which can be made by adding to the water
6 oz. of liquid ammonia, 1 of camphorated alcohol, and from
2 to 4 lbs. of common salt. On leaving this bath, the limbs
should be rubbed for twenty minutes with some camphorated
and stimulant liniment. Remak thinks highly of electricity
in these cases of hysterical paraplegia, one pole being applied
to the pit of the stomach, and the other to the lumbar region
of the spinal column, but this must be carefully done.

SCIATICA.—From four cases, I have selected one which
forcibly shows the utility of sudorifics in diseases of the
change of life.

CASE 41.—*Sciatica.*—Jane A., aged forty, had dark hair, grey eyes, small stature, nervous constitution, and was of slender make. After having suffered two years with violent headache and giddiness, she first menstruated between fifteen and sixteen, and continued regularly to do so very abundantly. She married at sixteen, became pregnant immediately ; had seven children ; was always regular until she quickened, and in three pregnancies she menstruated regularly up to the period of parturition. The previous summer she was treated for rheumatic fever, at the Royal Free Hospital. Menstruation proceeded regularly during the patient's stay at the hospital, appeared four times after her return home, and then stopped. She complained of an intolerable pain in the left lower limb ; and, on examination, I found that pressure on the spine did not increase or cause the pain. To all appearances, the left limb was as sound as the right ; and the pain was said to arise from behind the great trochanter, and to reach the back of the limb, following the course of the great sciatic nerve. The pulse was weak, the urine clear, and of the usual colour. I ordered a blister to the spot which was the most painful, pills of compound extract of colocynth, and a sedative mixture. The blister did not relieve the excruciating pain. Turpentine embrocations were ordered, the other measures continued, and the blister repeated, without the least benefit. March 8th.—The patient suggested that her pains might have been caused by the sudden cessation of menstruation five months previously. On learning that she had lately had flushes and nightly perspirations, never experienced before, I ordered her to take, as an emetic, 1 scruple of ipecacuanha with 1 gr. of tartar emetic, to continue the mixture and the pills, and to take a teaspoonful of flower of sulphur in milk on going to bed at night. 15th.—The heats and perspirations had increased ; the pains had much abated. I prescribed another emetic, and continued the other measures. 29th.—The pain is very trifling, and the heats, flushes, and perspirations, are more frequent, coming on, not only on exertion, but in bed, at night, and in the morning. April

Q

6th.—I ordered sulphur, 2 oz.; bicarbonate of soda, 4 drachms; and 2 scruples of ipecacuanha—a teaspoonful to be taken in a little milk every night. June 11th.—She has had no return of pain; but when the flushes and perspirations abate, she suffers either from headache or from the sensation of "something working" to bring on the pain in the leg. Jane A. has not suffered in any other way from cessation; and as she was subject to a copious discharge every month, one can understand that she should feel the effects of its sudden cessation; and as she was very nervous, it is not surprising that the nervous system felt the sudden shock, particularly that portion which was predisposed to disease by the previous attack of rheumatism. I saw this patient again in 1855; there had been no sciatica nor return of the menstrual flow, and her health was good.

Gardanne mentions, that a woman of a strong constitution, at her forty-fifth year, suddenly ceased to menstruate; and was at the same time seized with violent pains in the left thigh, and at the end of four months, was not able to move the limb. As she had suffered from syphilis in her youth, mercurials were given, but without effect. Sabatier and Gardanne then advised moxas to be applied to the leg, which produced slight fever and great perspiration, but restored the use of the patient's limb.

NERVOUS APHONIA.—This is a rare affection, but I have had a good opportunity for studying the case of a lady, at the change of life, who, after losing her husband, came to town, and settled in Belgravia. Though she had not hitherto been subject to nervous affections, cold, over exertion, worry, or sometimes no apparent cause, would suddenly deprive her of her voice for a few days. The nervous nature of the ailment was shown by the sudden coming and leaving of the aphonia, and by the effect of change of air; for a drive in the Regent's Park or to Hampstead would often restore her voice to its natural tone. She therefore left town to reside in the country, and has since enjoyed a comparative immunity from this complaint. Sometimes a potion, containing ether, speedily dispelled the aphonia. Dr. Deslieux has found chloroform useful, giving from 10 to 15 drops in a little water. In two cases, I found sudorifics useful, the permanent return of the voice coinciding with a marked determination to the skin. Cerise speaks in favour of emetics for nervous aphonia, and I have witnessed their sudden good effects, but the best treatment is undoubtedly the direct application of electro-magnetism either to the tongue or to the larynx by means of Dr. Morell Mackenzie's galvanizer. The shock makes the patient scream—the spell is broken, and the patient is immediately cured.

NEURALGIC AFFECTIONS OF THE EYES.—In two women, at the change of life, I have noticed the head symptoms to be accompanied by marked photophobia, and many patients complained of an unusual dimness of sight, which wore off on the abeyance of pseudo-narcotism, and other head symptoms. H. D., aged fifty, and chlorotic-looking, has been irregular for the last six months, and suffers much from intense debility, even after a full meal. She can scarcely keep herself awake during the day, but at night is kept awake by a dull, heavy pain in the eyes, lasting more than two hours, the probable cause of which being that she has lately worked too hard at embroidery. B. de Boismont gives the case of a woman who, at forty-five, was blind for three days ; Boyer rightly judged that it depended on the change of life, and on recovering her sight, she remained subject to giddiness. Dusourd met with three women who were blind for two or three days, at the change of life, and he has several times observed them affected with hemeralopia ; but Romberg erroneously considers women predisposed to amaurosis at the change of life.

DEAFNESS.—When deafness occurs at the change of life it should not be considered senile, for it may depend upon inflammatory affections of the internal ear, or upon morbid lesions of the external canal ; but deafness is, in general, purely nervous, attending on pseudo-narcotism ; the patients being, as it were, stunned, do not hear until they are fully roused from their state of torpor. I have seen this deafness appear and disappear suddenly; in one case it came on suddenly at cessation, when flooding abated about six years ago, but the deafness has continued ever since; and I have known women driven to desperation by continued noises in the ear, which baffle all treatment. M. S., a dispensary patient, aged fifty-two, never had a day's illness until forty-five, when she saw a man executed, which brought on flooding. The menstrual flow never reappeared, but loud and continued noises were constantly heard, which Mr. Harvey could not account for by a diseased condition of the internal ear. Cupping, blisters to the nape of the neck, and other treatment were useless. In a very nervous lady, erysipelas of the face occurred at the change of life, and left an excruciating pain, sometimes accompanied by a most annoying "forcing sensation" behind the ears. This had lasted for three years, but it yielded to general treatment, to the local measures necessary to cure uterine irritation, and to the rubbing in of belladonna ointment behind the ears.

RHEUMATIC PAINS.—Neuralgia of the nervous filaments of the skin is often caused by the damp state in which the underclothing is kept by continued perspirations at the change of life.

TREATMENT.—Heat under any form is good. Passing a hot iron over the painful part, previously protected by brown paper and flannel, may cure the patient. Vapour baths are serviceable. Sponging, with alternately hot and cold water, or shower baths of the same, will harden the surface against rheumatic influences. The thermal waters of Aix en Savoie are very effectual, and the best preventive is to wear flannel next the skin. In advanced age, these rheumatic cutaneous pains are often exceedingly troublesome, but women are then generally free from the eccentric nervous pains and temporary paralysis which have afflicted them in youth, and have been described, by Sir B. Brodie, as hysterical.

CHAPTER IX.

DISEASES OF THE REPRODUCTIVE ORGANS AT THE CHANGE OF LIFE.

TABLE XXVIII.

Liability to Diseases of the Reproductive Organs in 500 Women.

Flooding, a terminal	82
Floodings, successive.	56
Leucorrhœa, frequent, at irregular intervals .	146
„ monthly	12
Remittent menstruation.	33
Vaginitis.	4
Follicular inflammation of the vulva	10
Inflammation of the labia	4
Ulceration of the neck of the womb	9
Hypertrophic inflammation of the neck of the womb	2
Prolapsus of the womb	5
Uterine polypi	4
„ fibrous tumours	4
„ cancer	4
Chronic ovarian tumours	3
Irritation and swelling of the breasts . . .	14
Milky or glutinous secretion of the breasts .	2
Hard, non-malignant tumour of the breast .	2
Cancer of the breast	1
Habitual deposits in the urine.	49
Pain and difficulty in passing urine	9
Incontinence of urine	4
Hæmaturia	1
Erectile tumour of the meatus urinarius . .	2
Perineal abscess	2
Total	464

The foregoing table gives a fair idea of the derangements of the reproductive organs that were met with in 500 patients at the period of the change of life, various forms of suffering being frequently met with in the same patient. Hence it is clear that the change of life is a time of turbulent activity for the reproductive organs ; and though they are less liable than before to acute inflammation, they are more than usually so to congestion, hæmorrhage, mucous flows, and neuralgic affections.

When the change of life is over, the rule is, that the reproductive organs become more or less atrophied, and it is this process of involution which protects women against the frequent occurrence of inflammatory diseases of the womb. If these complaints occasionally occur, it depends on this process of involution having been checked by chronic inflammation ; in other words, uterine inflammation seldom occurs at, and after cessation, except in those who have previously suffered from it. After cessation uterine inflammation may be due to a fibroid development in the substance of the womb, or to a polypus growing from the cervical canal, or from the mouth of the womb. I sometimes find the os uteri ulcerated, but never to the same extent as at the previous period of life. When occurring at the change of life, it seems to me that uterine inflammation is oftener accompanied by biliary disturbance than at an earlier period, and I have known patients daily bring up what seemed a pint of bile, for two or three weeks, notwithstanding all remedies. Of sexual infirmities after cessation, it strikes me that vaginitis is the most frequent, sometimes originated and attended by slight ulceration of the cervix ; sometimes without any uterine lesion. The frequency of vaginitis is to be explained by the continuance of sexual intercourse long after cessation, for I have been repeatedly asked by women about sixty years of age whether, on account of vaginitis, it would be dangerous for their husbands to have connexion with them.

Sometimes vaginitis coincides with a similar affection of another mucous membrane, which implies a general failure of health, and explains the frequency of relapses. It may be laid down as a rule that, during the change and after cessation, inflammatory affections of the sexual organs are less frequent and less severe, and that when these organs then assume great activity, it is an activity of a lower type, denoting a deterioration of the plastic force; the ovary, the womb, the breast, becoming more frequently the seat of cystic, fibrous, or cancerous growths.

DISEASES OF MENSTRUATION AT THE CHANGE OF LIFE.— Flooding seems to have been the only disease of menstruation, at this period, which has engaged the attention of medical men; but besides flooding there is the *stillicidium uteri*, or continual dribbling from the womb, lasting for weeks or months, and effectually undermining strength. The pains accompanying menstruation may become unusually intense, and have been considered in the chapter on "diseases of the nerves;" but the mode of recurrence of the menstrual flow calls for some remarks. It is better for women that the menstrual flow should cease gradually, and its stopping suddenly may depend upon some removable circumstance. Attention to the general health, sedatives, mild tonics, and purgatives, an occasional tepid bath, and rest will often cause the flow to reappear with comparative regularity, for some time longer; but I do not agree with Dr. Ashwell in recommending mustard hip-baths, stimulating embrocations, and sexual intercourse. Instead of returning every two or three months, the menstrual flow may return every two or three weeks. This occurred in twenty-six cases out of 383, in whom the catamenia were ceasing. It is unnecessary to point out how weakening and how irritating to the nervous system is this too frequent recurrence, which can, in most cases, be prevented by the exhibition of sulphate of quinia, in three-grain doses, to be given at night. If quinia, combined with the avoidance of hot drinks and rest do not check the too frequent flow, I give a mixture with sulphuric acid and alum, and advise the use of cold vinegar and water to the vulva and adjacent parts.

FLOODING.—Mauriceau, Levy, and others, look upon flooding, at the change of life, as indicating cancer of the womb;

Astruc, Gardanne, and some modern pathologists, consider it
to depend on ulceration of the womb or its chronic hypertrophy; but as I have known women, free from all uterine
symptoms, and in every other respect well, have floodings
from a fall, from suddenly hearing of the death of a relative,
flooding need not be accounted for by ulceration of the neck
of the womb, and to test the accuracy of that opinion, I
examined twenty such patients, and in two only did I find
uterine ulceration. This brings me again to the diagnosis of
flooding at the change of life, for I have already shown, at
p. 40, that the flooding of internal metritis, uterine hydatids
and fibroids, uterine polypi and cancer, are sometimes attributed to the change of life. I have known these mistakes to
be made by the most talented, because they would not take
the trouble to make a digital examination before giving a
diagnosis. I have thus seen women reduced to the last
stage of debility by a uterine blood flow, more or less
abundant, being allowed to continue for two, three, or four
years without an examination being made, because these
patients were at the time of life when the menstrual flow
usually ceases. Quinia and port wine were given, and
change of air recommended; but no astringent vaginal injections, not even cold applications, had been made. I have
now under my care a lady, with cheeks as white as my paper,
who has been flooding more or less for the last two years, and
who was repeatedly told there was nothing to be done, that
it depended on the change of life. On examination, I found
a polypus hanging to a cavity of the womb by a pedicle as
thick as my little finger. Another lady was allowed to flood,
at short intervals for three years, and on examining her I
found the cervix three or four times the right size, giving the
idea of a soft and podgy substance, and quite denuded of
epithelium. My first impression of the case was cancer, but
I afterwards became convinced it was not specific, for I cured
the patient by the means of nitrate of silver; and she regularly
menstruated for several years.

When flooding occurs at a catamenial period, as a critical
discharge, subsiding completely after a few days, every one

recognises it as depending on the change of life; likewise if successive floodings occur at successive menstrual periods. When flooding occurs a few months or even years after the absence of the menstrual flow, and is accompanied by flushes, perspirations, and other symptoms of cessation, without being explained by uterine disease, it is fair to suppose that it depends upon the change of life; but whenever it does not appear as a downright critical flow, when it is followed by a lingering sanguinolent discharge, with returns of flooding at irregular periods, with leucorrhœa and other uterine symptoms of disease, the case is doubtful, and no medical man can conscientiously give a diagnosis without a previous examination. Who has not been consulted, for flooding, by women of forty-five in every appearance of health, and has not been grieved to find such cancerous lesions of the womb as warranted the conviction that the patient's days were irrevocably numbered? Louis and Valleix have mentioned that, before any visible cancerous changes have taken place in the womb, profuse menstruation will sometimes be the first symptom of the complaint. This is an additional reason why profuse menstruation should be frequent at cessation. No other cause of flooding is wanted besides the great predisposing cause already discussed; but it may sometimes be occasioned by a fall, a fright, a violent fit of anger, by sneezing, or by connexion. Fothergill has seen an intermittent fever at cessation give rise to menorrhagia every month.

Beyond a certain loss of blood, which must be estimated by the patient's pulse, by the expression of her features, by her feelings of strength or weakness, the further flow is dangerous. If those who, at this period, lose too large quantities of blood, do not quickly repair the loss, they can seldom do so; and, although not completely invalided, they remain pale and chlorotic, they do not enjoy the degree of strength with which they were previously gifted, and become nervous. There is, however, a certain amount of danger in stopping the critical flow too soon, particularly in those who are plethoric, or who have been accustomed to lose much blood at the menstrual periods. J. Frank says he has often seen

apoplexy brought on by means injudiciously used to stop the
floodings of cessation ; and two cases in my own practice
confirm this remark. F. Hoffman has seen apoplexy at the
change of life in women subject to an abundant flow. This
sufficiently shows that it is wrong to let flooding at the
change of life take its chance, and that it should be considered
a complaint requiring medical advice. The treatment of
flooding during the attack should be directed so as to check
the flow, by placing the patient in the horizontal position on
a horse-hair mattress, with light covering, in a cool room ;
by giving cold lemonade made with the *mineral* acids, alum
whey, ergot of rye and strychnia, or nauseating doses of anti-
mony. Local measures may be required, such as iced
vinegar and water to the abdomen, and to the inner parts of
the thighs; or a lump of ice to the neck of the womb, and
by making her grasp a lump of ice, for the cold thus transmitted to the nerves of the womb may suffice to stop the
flow. The injection into the vagina of a strong solution of
tannin, or alum and zinc, has been sometimes successful.
Dr. West has shown the utility of intra-uterine injections of
a solution of gallic acid, or an infusion of matico, to stop
continued flooding after all other means had failed, in a woman
of fifty-one, and I have found a strong solution of the perchloride of iron still more useful. I have thus, even in cases
of cancer, checked bleeding that had been going on for more
than a year, but in a doubtful case that I saw, with Mr. Hol-
berton, the injection caused such tenesmus, and so great an
amount of pain continued for several weeks, that I asked
myself whether the remedy was not worse than the disease.
Caseaux has seen many women die several hours after puerperal hæmorrhage had been stopped, too little blood having
been left to stimulate the brain and nervous system, so as to
enable it to perform the indispensable vital acts of respiration
and circulation. In such cases he recommends circumscribing
the blood into the smallest possible space, by bandaging the
four limbs, and by pressure to the aorta. Similar measures
might be useful at the change of life in exceptional cases.

It has been pointed out, that in the plethoric and those accustomed to menstruate profusely, bleeding often did good at the time, and prevented subsequent ill-health; but it is seldom advisable to bleed the month following a flooding, or the one after that, unless there be evident signs of congestion. It was for continued flooding that Fothergill, Hufeland, Lisfranc, recommended the plan of taking less and less blood, from a patient, at successively longer periods, and the practice is sometimes very useful. By bleeding we take away the material of flooding; but it must not be forgotten that hæmorrhage as often depends upon the perturbed action of the blood-vessels as on plethora; and although other measures may be indispensable, the return of many hæmorrhages may require a judicious use of sedatives. A sedative mixture, containing Battley's solution, or a solution of acetate of morphia, taken at night, will therefore be found useful; and as the blood is directed with such force to the womb, it becomes a centre of morbid attraction requiring to be lulled and stupefied. This may be done by giving sedative injections by the rectum, or a suppository made of one grain of opium and two of extract of belladonna, may be used every night, until the nervous irritability of the reproductive apparatus be quelled. Saline purgatives and small doses of nitre are also indicated; and it should be borne in mind that at the dodging time, or after cessation, no centrifugal tendencies of blood should be encouraged by pediluvia, hip-baths, mustard poultices, or by similar applications to the lower extremities.

LEUCORRHŒA, or a flow of mucus unmixed with pus, at the change of life, is, like flooding, a critical discharge, the indiscreet interference with which might cause a more serious complaint, and I have met with those in whom it was immediately increased by wine and by emotion. If the discharge becomes abundant the best way is, not to check, but to regulate it, by increasing the solubility of the bowels and the habitual moisture of the skin. Frequent lotions with tepid water may be recommended, but I very rarely order

cold injections, and, when these are necessary, seldom with
the addition of any astringent substance; after a time,
injections of water, containing 1 drachm of acetate of lead
or of alum to the pint, are useful.

I have met with women who have a continual sero-sanguinolent discharge of no serious import. In one case the
lady was remarkably strong and healthy, and menstruated
with great regularity up to her death, at eighty-four. Six
years before this lady's death I was consulted, and I found
a considerable enlargement of the capillaries of the vagina,
particularly on the cervix, but no redness of the mucous
membrane itself, and no ulceration, enlargement, or cancer.
Several applications of the solution of nitrate of silver did
no good, neither did injections, so all treatment was aban-
doned. A very strong lady ceased to menstruate at fifty,
and at fifty-two she was suddenly told that she was to be
deserted by a man who had lived with her as her husband
for twenty-five years. The shock brought on delirium which
lasted a fortnight, and a muco-sanguinolent discharge from
the womb, which has continued ever since, and has now
lasted three years, without impairment of health. There
was neither ulceration, polypus, nor cancer to explain
this discharge, which came from the internal surface of the
womb, only inconvenienced the patient, and was not much
diminished by astringent injections. Another patient has a
pale pink discharge whenever she strains, or remains standing
for long, and whenever she puts her hands into warm water.

Vaginitis.—The occurrence of vaginitis at the change of
life or afterwards was only noted four times in 500 patients,
but I have seen many cases of this complaint during the
last twelve years, and sometimes the patients were mothers
of large families, and had never suffered till cessation, when
vaginitis became troublesome. Vaginitis is often attended
by great internal heat, bearing-down pains, vesical disturbance, with or without scalding sensations on passing urine.
The discharge is sometimes purulent, offensive to the nose,
and acrid enough to chafe the pudenda. At other times it

is milky or watery, and the gentlest digital examination is most painful. Before ordering vaginal injections, in very acute cases, it is better to reduce the inflammation by a saline purgative, and copious dilution with imperial drink, or linseed tea; by large, thin, warm, linseed-meal poultices applied to the lower half of the abdomen; by lotions, with linseed tea, adding to each pint one or two drachms of acetate of lead; by the rectal sedative injections; and by a tepid bath, prolonged for an hour. In a few days the lotion may be used as a vaginal injection, and it may be requisite to apply to the vagina, every fifth or sixth day, a solution of nitrate of silver, forty grains to the ounce of distilled water.

It is not unusual for those who have suffered much from uterine disease to have occasional attacks of vaginitis after the change of life, and four patients have for several years been obliged to come to me on this account, once or twice a year. In them there is no trace of uterine disease, but sometimes vaginitis coincides with an enlarged and displaced womb, or with a fibroid. I have just seen a lady who, six years after the ménopause, was given up as suffering from cancer, and in whom I can only find extensive vaginitis, with a large anteverted womb. Sometimes vaginitis seemed to be the result of a tendency to mal-nutrition of most of the mucous membranes, sometimes I traced it to too frequent connexion in advanced life. I have also seen acute vaginitis in married women in whom there had never been intromission, and I have met with women who had previously had children, and nevertheless inflammation and spasm had so contracted the vagina that I could with difficulty introduce the forefinger. Occasionally vaginitis is the only organic lesion that accompanies distressing cases of neuralgia of the sexual organs, to be hereafter considered, but regarding which I may already state that the neuralgia persists after the cure of the vaginitis.

FOLLICULAR INFLAMMATION OF THE PUDENDA.—This form of disease is most frequent during previous years, but I have occasionally known it to begin at the change of life, and to be subject to occasional relapses for the following three or four years, while I have met with three instances in which the disease had begun about thirty, and continued many years after the ménopause. The annoyance suffered is extreme, and the irritation such as to render the application of the hands imperative, and causes the discharge, at other times mucous, to become bloody. In one lady the irritation came on periodically, and then several hard lumps, about the size of a filbert, would appear in the margin of the labia. These swellings seldom suppurated, but occasionally vesicles appear on the skin. I have observed that in severe cases of this description, there is no exaltation of sexual desires, connexion being on the contrary abhorrent to the feelings, though it may be otherwise when the affection is slight.

The labia should be fomented every two or three hours, with a lotion containing half an ounce of acetate of lead, and two drachms of laudanum to four ounces of distilled water, and fine linen should be steeped in the lotion, and gently applied to the irritated surface, using afterwards a powder made of powdered starch, ʒv ; powdered acetate of lead, ʒss ; and essential oil of bitter almonds, ℳx. If that does not do good, I recommend a lotion made with two drachms of sulphate of zinc, two drachms of diluted hydrocyanic acid, and two ounces of glycerine to sufficient distilled water to make an eight-ounce lotion.

A tepid bath, or hip bath, should be taken daily, or every other day, warm water being added, so that the patient may remain in it for an hour, or more, if possible. After the full effect of a saline purgative, a sedative rectal injection should be given once or twice in the day ; and at night a scruple of Dover's powder, or of Battley's solution, may be beneficial. Should this plan of treatment fail, it would be necessary, instead of the acetate of lead, to have recourse to the application of a solution of nitrate of silver. The patient cannot effectually apply this remedy, which must be done by the medical adviser, who will have to steep cotton wool in a

solution of nitrate of silver—40 grains to the ounce of distilled water—and to rub it well into the recesses of the
mucous membrane, before its morbid excitement will abate.
Trousseau recommends, as a never-failing lotion for this
affection, a pint of water, in which is dissolved a large pinch
of a powder made with equal portions of deutochloride of
mercury and muriate of ammonia. He also recommends
first two vaginal injections a day, then one of a solution of
deutochloride of mercury, ʒj to the pint of cold water.
I agree with Huguier, that the vulvo-vaginal glands are not
liable to inflammation after forty-five.

PRURIGO PUDENDI.—This is a common form of prurigo,
and after cessation it is often accompanied by prurigo
of some other part of the skin, or prurigo senilis. This
is a most distressing disease, and Lorry asserts that the
mere local irritation is often succeeded by a fearful desire
for sexual gratification. I have seldom met with severe
cases, but B. de Boismont, Gilbert, and Mr. T. Hunt, note
prurigo and eczema of the vulva and anus as not unfrequent
at the change of life. The following case is derived from
Mr. T. Hunt's practice.

CASE 42.—*Herpetic affection of the labia.*—A single lady of
fifty, who had never menstruated, suffered from irritation,
chiefly confined to the left side of the clitoris, which was in
a state of hypertrophy from the constant friction. As the
patient could not well retain her water, a careful examination was made, when it was found that the meatus was larger
than usual, and that *there was no vagina or uterus.* When
young she frequently had leeches applied to the labia to
induce menstruation, but no examination had ever been
made with a view of ascertaining the cause of the amenorrhœa, nor was she at all aware, up to the age of fifty, that
she was the subject of malformation. She had all the other
physical characters of a female, together with the modesty
and delicacy of mind peculiar to her sex; there was sexual
passion, and she would long previously have been married,
had she not feared that matrimonial unhappiness might arise

out of the total absence of the catamenia. She was never conscious of any local pain or fulness at the monthly periods; but she was frequently troubled with headache and giddiness at irregular intervals. In other respects her health was excellent; but the pruritus, which was often accompanied with an herpetic eruption on one of the nymphæ, was the great plague of her life, and strong religious principles had alone prevented her committing suicide. Arsenic and other remedies were tried in vain.

Mr. Hunt speaks most favourably of arsenic. In milder cases, saline purgatives have proved very useful; also, prolonged tepid baths, cold baths, cold hip baths, and cold water injections. The lotions recommended for vulvitis will be equally good in prurigo. Ointments with a preparation of zinc, lead, or mercury, may be tried, as well as powders containing calomel and camphor. I have given camphor in various ways, with good effect. Lotions made with the sedative camphorated lotion, or with camphorated vinegar diluted with water, camphor ointments, camphor powders, and camphor sprinkled between the bed and sheet. The utility of preparations containing carbolic acid confirms Hebra's assertion that prurigo is caused by pediculi, and suggests the careful cleansing of the patient's clothes so as to destroy the insects by which the complaint is often indefinitely prolonged.

NEURALGIC AFFECTIONS OF THE SEXUAL ORGANS.—In the first place, I must treat of Erotomania, or the inordinate desire for sexual gratification, which is a mental phenomenon, and should have been therefore placed on the list of irresistible impulses, in the chapter on cerebral affections. Pudendal neuralgia will then claim the reader's attention; and lastly, I will describe as ovario-uterine, a little understood form of abdominal neuralgia.

Erotomania.—The inordinate desire for sexual gratification, described by writers as *furor uterinus, uteromania,* and *nymphomania,* is a mental affection, suggested, promoted, and intensified in a very limited number of women by morbid ovarian influence, by uterine affections, and by the various kinds of pudendal disease that I have just described; and this sufficiently explains the importance of curing them.

The Greeks considered erotomania to be the result of Divine vengeance for neglecting the worship of Venus. In the Middle Ages, the complaint was considered to be a proof of Satanic influence; but we now rightly place in lunatic asylums those who suffer from it to an aggravated extent. I have stated at page 111, that the most distressing cases of crotomania are very rare at the change of life; but B. de Boismont had six cases in his asylum. Mathieu cites one case in a lady at fifty, and another in a woman at sixty years of age. Dusourd mentions three, and Louyer Villermay cites two cases, as having occurred after cessation. I do not agree with Mathieu and Roubaud that the disease is frequent at the change of life, for I have not met with a single instance of an aggravated nature. There is, however, so much of truth in the old French proverb about " Ce diable de quarante ans si habille à tourmenter les femmes," that towards the change of life, there is often a temporary rekindling of passion thought to be altogether extinguished, like the sudden re-awakening of a smothered flame. This explains why, at the change of life, women sometimes take a husband when it is treatment they require, and why, occasionally a married woman who had hitherto led so virtuous a life that it seemed impossible for her to go wrong, will actually sacrifice position, friends, children and husband to begin a new life with a worthless vagabond. With regard to treatment I will only say that it behoves the medical adviser to remove any kind of morbid irritation that may exist, not only in the pudenda, but in the womb and ovaries, and those who have to treat severe cases of crotomania in lunatic asylums must remember that a digital examination will not enable them to decide that there is no uterine disease; that point must be determined by an instrumental examination, the patient having been previously placed under the influence of chloroform; indeed, it seems to me

R

that medical men are incapable of well attending insane women, unless they thoroughly understand diseases of women.

Pudendal neuralgia.—It is easily understood that morbid lesions of the pudenda should give rise to more or less heat, aching, or itching ; but occasionally at the change of life one or other of these sensations assume a very distressing and long continued intensity, although no structural lesion of the pudendal tissues can be detected. Many patients complain that for a few nights after each menstrual period, there is an intense itching and skin irritation, unaccounted for by any cutaneous eruption. Such cases explain why cessation is sometimes followed by itching of the pudenda, or of the whole surface,—the prurigo latens of Alibert, to which Aran believed women at the change of life to be particularly subject, and to indicate some form of uterine disease. In some of the most distressing cases of pruritus and perverted feelings, there is, however, nothing morbid to be detected, as in a woman of sixty, who was subject to hysterical attacks if a young man came near her, and in whom Biett could discover nothing amiss, although he inspected the pudenda with a magnifying glass. Lisfranc and Tanchou have drawn attention to vulvo-vaginal neuralgia ; and the last thinks it occurs most frequently at the change of life. He mentions the case of a lady of forty-eight, who had never suffered from uterine disease, in whom the menstrual flow ceased at forty-six, and ever since she had been afflicted with pruritus and darting pains in the pudenda. These sensations were increased by walking and over-exertion ; no morbid lesions could be found to explain the sufferings ; and they were relieved by the internal exhibition of assafœtida and valerian, combined with the external application of opiates and belladonna. It must be confessed, however, that some of these cases stubbornly resist remedies, and yield only to those influences that have power to strengthen the nervous system, and to those nutritial transmutations of tissue that time alone brings about.

Ovario-uterine neuralgia.—I am not aware that cases similar to mine have been described by gynæcologists, but Dr. Handfield Jones has recorded two similar cases in his work on functional nervous diseases. One of his patients was thirty-eight years of age, the other younger still, but four of my cases occurred at the change of life, and two after cessation. These cases were characterized by a most distressing burning sensation in the vagina and pelvis, with a considerable amount of tenesmic action, shifting from the rectum to the bladder or to the womb, and sometimes felt in the three organs at once, the sensation being sometimes so well localized as to impress the patient with the conviction that there was a solid body pressing on the rectum or the bladder, or a tumour on the womb. These pelvic symptoms are said to bring on exhaustion and prostration referred to the pit of the stomach, followed by sleeplessness or pseudo-narcotism, despondency, apathy, and other forms of cerebral neuralgia described in a previous chapter, as gangliopathy and ganglionic dysæsthesia, for these pathological conditions are closely related to that form of neuralgia of both sensory and ganglionic pelvic nerves which I think right to call ovario-uterine. I am strengthened in the belief that the ovaries are greatly implicated in this neuralgia, by my having observed a somewhat similar train of pelvic symptoms to occur in man as the result of concussion of the testicle. Very nervous women are most liable to the complaint, but I have seen it assume its worst form in strong-minded women, who, through life, had always enjoyed good health. I have always found the neuralgia accompanied by a certain amount of vaginitis, and its cure always alleviated the symptoms, but seldom cured it. Vaginal injections were often useful, but sometimes an injection brought on a fit of neuralgia, and for a time injections had to be left off. Worry, mental anxiety, and fatigue made the symptoms worse. A tranquil tenor of life, rest, particularly in the recumbent position, with the feet higher than the pelvis, quieted the symptoms for a time; and they sometimes yielded speedily to opium and belladonna suppo-

sitories introduced into the vagina, or the rectum, but the
cure of the complaint must be sought for in whatever · can
strengthen the system, such as change of scene, travelling,
more cheerful circumstances, the exhibition of our ordinary
tonics and those called nervine, like arsenic and strychnia.
Although I have only met with six aggravated cases of this
description, I believe that in a less degree the complaint is
far from uncommon, remaining unnoticed, because many
women effectually conceal their sufferings. The following
cases give a good idea of the disease :—

CASE 43.—*Ovario-uterine neuralgia.*—Miss X., was forty-
seven when she first consulted me. She is small but well
proportioned, has been highly nervous all her life, and is
so still. Menstruation was irregular, and there was a muco-
purulent discharge, vaginitis, and decided ulceration of the
cervix, and a most irksome sensation of heat and irritation
in the passage. I cured the vaginitis and ulceration by
surgical measures without relieving the vaginal heat and
pruritus, so I sent the patient out of town. When she
returned after many months, the pruritus was as bad as
ever, and would come on after any excitement or fatigue
or standing about, and would be relieved by resting with
the feet higher than the pelvis.

This vulvo-vaginal irritation would sometimes disappear on
the coming on of a similar pruritus, on the palms of the hands
and on the soles of the feet, showing that however much the
chief seat of neuralgia might be in the womb and vagina,
the extreme nervous expansions in other parts of the body
might suffer in like way. When this irritation affects the
feet and hands there is nothing to be seen there, and she
refrains from scratching them, because it would make the
irritation last for hours. As might have been predicated, the
symptoms are worse at night, and lead to great exhaustion and
despondency. I have watched this state of things for ten years
and often could give no relief. She was always better for plenty
of food and wine, and for such small quantities of citrate
of iron and quinia as she could bear. I tried all sorts of

injections, tar water did most good, but it has been repeatedly advisable to leave off all kinds of injection, for they
seemed, for a time, to do more harm than good. I syringed
the vagina with a solution of nitrate of silver, and touched
the passage with the solid caustic with questionable benefit.
A rectal suppository containing a grain of opium and one
of extract of belladonna often gave temporary relief, but
this remedy could not be relied on. Many a daughter has,
by the sacrifice of her own health, repaid her mother, the
gift of life, and a year ago, my patient lost a mother who
had been long a cripple, requiring anxious and fatiguing
nursing; and afterwards she went out of town and got fat,
and now suffers much less, having a slight return of the
old symptoms, when she gets weaker and more nervous.

CASE 44. — *Ovario-uterine neuralgia.* — A very strongly
constituted lady, aged forty-seven, is said to have had
some acute uterine disease twenty years ago, while residing in France, when forty leeches were applied above
the pubis. With the exception of not being able to
retain the urine so well as previously to this attack, health
remained so good that every year she was able to take
long pedestrian excursions with her husband. She never
conceived, and menstruation ceased suddenly at forty-four;
for the following months the nose bled very frequently
and the bowels became constipated; for this she went to
Homburg, and was restored to health. On returning to town,
in December, 1868, she took very cold enemata for constipation, which was so great that a wineglass of Friedrichshall
water, taken every hour, failed to produce watery motions, but
irritated the bladder, and was followed by anomalous abdominal sensations, that have lasted ever since. She feels as if
there were a heavy body in the pelvis, bearing down upon the
rectum, with a burning sensation, referred sometimes to that
organ, sometimes to the vagina, or to the bladder. When
in bed and lying down, with the feet up, the patient feels
comfortable; by the time she has half done dressing, the
burning sensation begins, and lasts until the bowels have

been moved; soon after this the burning comes back; it is aggravated by standing or sitting, by indigestion, flatulence, constipation, and repletion of the bladder; also by worry and bad news. The sensation is relieved by walking, by lying down, and by regularity of the bowels. Homburg was again tried, and did good, but on her return the lady was as bad as before, and consulted several doctors. One attributed the sufferings to stricture of the rectum, another to irritation of the bladder, a third to displacement of the womb. In due course of time Homburg was tried for a third time, but the waters were soon left off, as they aggravated all the symptoms, and after the patient's return to town Dr. Beale sent her to me. In addition to the pelvic symptoms already described, there was considerable cerebral disturbance, for a strong-minded, sharp, matter-of-fact woman, was in a state of mental confusion, her brain felt muddled, and she would sit for hours dozing or doing nothing; despondency being doubtless increased by finding herself helpless as a child, after having passed all her life in doing everybody else's business as well as her own. She forgets where she puts things, once thought she had taken out a large sum of money in her purse, and that she had lost it, whereas a month afterwards she found it in some out of the way place. On examining, I found the rectum perfectly healthy, notwithstanding the pain and stricture ascribed to it. The vagina was so narrow that I could with difficulty introduce part of my index finger, and I gave so much pain that I desisted from all further investigation, ordered linseed tea and laudanum injections, three times a day, and henbane internally. A few days afterwards I was able to reach the os uteri with the tip of my finger; I found the womb exquisitely sensitive; and wishing to ascertain how far the bladder was implicated, I sounded it, found nothing abnormal except great pain when the sound passed over the urethra, but the pain was not caused by inflammation, for the finger in the vagina did not enable me to feel the urethra as a hard and round body, giving pain on

pressure. Injections with acetate of lead and laudanum, as
well as opium and belladonna rectal suppositories, enabled
me a little later to examine the womb without giving pain;
there was no ulceration, and all along there had been very
little vaginal discharge. The pain was most felt at the
opening of the vagina, which looked sore, red, and injected,
a condition that may account for an unusual hardness of the
intervagino-rectal tissues, a hardness of which the patient is
sensible, and complains of as something wrong with " the
bridge." This is caused by long continued congestion,
and the parts are now without heat or redness. This
sore state of the vaginal opening was relieved by the
application, twice a day, of zinc ointment, to each ounce
of which was added a drachm of diluted hydrocyanic
acid. Vaginitis becoming worse, I swabbed the vagina,
once a week, with a solution of nitrate of silver, and I
ordered alum and zinc injections ; but suppositories, whether
administered by the vagina or the rectum, did harm. After
thus treating the patient for a few months, the sensations of
burning and weight had considerably diminished, but were
often still very troublesome. Digestion was much improved
by nitro-muriatic acid and pepsine, pseudo-narcotism and
mental disturbance were not relieved by bromide of potassium,
but were much reduced by henbane and Indian hemp, and
for the last six weeks the patient has been taking, at three
meals, the twenty-fourth of a grain of arseniate of iron,
made into a pill with the eighth of a grain of Indian
hemp; a combination suitable alike to the general nervous
derangement and to the abdominal neuralgia. This leads
me to the question of diagnosis. There was no organic
disease of the bladder or rectum, nor of the womb, neither
displacement nor ulceration of this organ. The disease was
vaginitis, kept up by excessive walking at the change of
life. The vaginitis causing neuralgia of both the sensory
and the ganglionic pelvic nerves, the vaginitis and neuralgia
causing pseudo-narcotism and the other forms of cerebral
disturbance that usually attend the ménopause; the neuralgic

element of the case being shown by the patient's often feeling
the disturbance to ascend, as it were, from the pelvis along
the spinal column, to the back part of the head, where there
is most suffering. A residence at the seaside, and the continued use of injections and internal exhibition of strychnia
will, I believe, complete the cure.

DISEASES OF THE CERVIX.—Hard hypertrophy of the cervix
has been admitted to occur at the change of life by Gardanne
and B. de Boismont, but whenever I have met with cervical
enlargement at the ménopause, it was evidently the legacy
of bygone years. Hard hypertrophy may then last for
a long time, with slight symptoms or none, and I have
never seen it turn to cancer.

I have only twice met with soft hypertrophy or engorge-
ment originating at the change of life, and it has been
observed by Dr. Gunning Bedford and by Dr. Forget. The
cervix was much enlarged, felt soft and boggy, as if constituted by erectile tissue. Examined through the speculum,
the cervix presented a ragged wound which bled freely. In
this form of disease, the repeated and abundant flooding
principally comes from the cervix, and not from the cavity
of the womb. My two cases had been taken for cancer,
but the movability of the womb, the healthy appearance
of the vagina, the absence of severe pain, and of the
cnaracteristic smell of cancer, enabled me to establish that
it was not specific. Forget cured his case by repeated
cauterization, and I mine by the alternate application of
the solid nitrate of silver and its solution.

Ulceration of the os uteri.—It must be borne in mind that
not unfrequently women have been suffering from undetected
ulceration of the womb, for a variable period of time, when
the sufferings of the ménopause cause it to be detected ; and
while this sheet is passing through the press, I have been
consulted by a very strongly constituted widow, who has
evidently been suffering from inflammation of the cervix
ever since her last confinement, twenty years ago. The
gradual ceasing of menstruation increased her suffering,

and I found an ulcer about the size of a shilling at the os uteri, dipping into the cervix. Other women who had been well cured of uterine ulceration, many years previous to the change of life, then have relapses, and I have known some have return of this disease four years after cessation. A patient, now 60, had a first and severe attack at cessation, and has a slight return every year. Another who is married, but without children, had a first attack four years after cessation fifteen years ago, and ever since there is a tendency to vaginitis. When ulceration occurs after cessation there is frequently an hypertrophied womb, or a tendency to vaginitis, and I have repeatedly seen the point of insertion of small polypi to be the centre of a patch of ulceration varying in extent from the size of a shilling to that of a florin, and, in four cases, the inflammation and ulceration of the cervix seemed to have been caused by marriage during the dodging time.

I avoid repeating here, what I have written on the treatment of uterine inflammation, in my " Handbook of Uterine Therapeutics," but I will briefly state that, if the case be one of inflammation of the os uteri with erosions, it would be well freely to paint the diseased surface with the solid nitrate of silver, and after this has been repeated once, or more frequently, at five days' interval, a solution of the same salt might be applied. I have tried alcohol, vinum opii, chromic acid, carbolic acid diluted with glycerine, the styptic colloid, but I prefer the nitrate of silver, for it is the best mucous membrane improver with which I am acquainted. It is, however, well to have a change of topics, for whenever uterine ulceration occurs in advanced life it is much more difficult to cure, as Dr. Bennet has already stated. In all such cases vaginal injections should be made by the patient two or three times a day, with linseed tea; a solution of acetate of lead, or of alum, one or two drachms to the pint. The patient, using an india-rubber syphon syringe, lying down when using the injection, and prolonging its use for five minutes at least. It will sometimes, however, occur that injections do no good, and cause pains like those of impending menstruation, and such cases have made Dr. Marion Sims too much disparage injections. I remember two cases in which the mouth of an atrophied womb was surrounded by a distinct rim of ulceration, which I could not remove by this plan of treatment, nor by the application of the acid nitrate of mercury or potassa fusa c. calce, and I was obliged to content myself with keeping in check what I could not cure, though time may.

INTERNAL METRITIS.—Even in a chronic form I consider internal metritis to be a very rare disease at, or after the change of life, and I have only seen one case of acute internal metritis after the ménopause which deserves to be recorded.

CASE 45.—Mrs. T., aged sixty-four, was sent to me in 1868 by Dr. Smith, of Weymouth. She was very stout and florid, and it appears that she miscarried several times when twenty-five, and that the late Dr. Lever told her that if she did not take more care of herself she would suffer for it later in life. When forty-four she consulted Dr. Smith, who has thus kindly informed me of what occurred when the patient was under his judicious care :

" I first saw Mrs. T. in 1847, for menorrhagia and hysteria. The womb was greatly congested, and I applied leeches to it several times, and sent her home apparently cured in Sept. 1847. She returned to me, however, in Sept. 1850, with similar symptoms, and was a month under my care, with no other treatment than saline aperients and astringent vaginal injections, and she went home as before apparently well."

Menstruation ceased at fifty without bad symptoms, but when fifty-four she began to lose blood from the womb, and continued to do so more or less until the day of her death. As soon as menorrhagia set in, Mrs. T. went to Weymouth, and with reference to this period Dr. Smith writes me word:

" I heard nothing of her again until Sept. 1867, when she

returned complaining of a sense of weight in, and bearing
down of the uterus, with occasional discharge of blood, often
attended with exquisite pain, vaguely described as being at
one time in the uterus, again in the bowels, back, and hips.
The uterine sound passed three and a half inches into the
uterus, and did not give more than ordinary pain. I failed
to check the bleeding, except for a time, though she was
about seven weeks under my care. I dilated the os with
sponge tents and tangle with the view of reaching any morbid
growth, if any, in the uterus, but found none. I came to the
conclusion the case must be one of fibroid tumour in the
uterine wall, since all the ordinary mediums to check hæmorrhage and allay pain had failed."

When I saw Mrs. T., in September, 1857, I found the
womb about double its usual size, and painful when pressed ;
the cervix was much enlarged, and not at all ulcerated. I
easily introduced three inches and a half of the uterine sound,
and this gave great pain. If the flow of blood was checked
for a limited time, a little pus would be passed in a gush and
after great pain, leading me to believe that it had accumulated
in the womb ; but independently of the passing of blood or
matter, Mrs. T. complained most of attacks of pelvic forcing
pain, lasting from one to two hours. When she got worse, I
saw her in one of these attacks, and nothing more resembled
the pains of labour, as she lay on her back, with face injected,
groaning or screaming, and tugging hard at a sheet tied to
one of the bedposts. These pains increased, the paroxysms
became longer, brought on vomiting, and they prevented
sleep by coming on at night. The sickness was for a time
relieved by an effervescing draught with prussic acid, but the
various sedatives that I tried internally, externally, and by
the hypodermic method, muddled the brain without easing
the pain or inducing sleep. Sulphuric acid, tannin, and ergot
of rye did not check the hæmorrhage, nor did various cold injections. I could not persuade Mrs. T. to let me inject the
cavity of the womb with a solution of perchloride of iron
until the plan was sanctioned by my friend Dr. Barnes. I

easily injected from 2 to 3 ounces of equal parts of the
strong perchloride and water, which caused very great faintness, but no increase of pains and marked diminution of
blood loss. I repeated a similar injection a week afterwards,
causing great aggravation of pain, which continued un-
ceasingly; she gradually took less and less food, slept less
and less, and died of exhaustion on the 17th of April, 1868.
Notwithstanding my respect for Dr. Smith's opinion, neither
I nor Dr. Barnes can see sufficient reason to suppose that
this fearful pain was due to a fibroid developed in the
substance of the womb. The paroxysms of pain were
those of most acute metritis, and the tablespoonful of pus
that gushed from the womb after increased pain could
only have been generated in its internal cavity; but I
could not prevail on the relatives to let me settle the point
by a post-mortem examination. I cannot conclude the case
without stating that I believe it would have had a fortunate
termination if I had been able to persuade the patient to let
me inject the womb at a much earlier period with a solution
of perchloride of iron, or of nitrate of silver.

Before commenting on chronic internal metritis, I will
state that I have never known uterine exfoliation to occur
towards the ménopause, although singularly enough, in the
first recorded case of uterine exfoliation, Morgagni mentions
its continuing up to cessation, and Dusourd has met with
similar cases. Those bodies called *fongosités* by Recamier
being evidences of morbid action on the part of the uterine
mucous membrane, it is easy to understand they may sometimes account for flooding at cessation; but it must also be
remembered that the same bodies have been found in women
who had no red discharge from the womb.

Chronic internal metritis.—During the last ten years I have
seen three cases, in which, during "the dodging time," the
patients suddenly passed more or less fetid pus at menstrual
periods, after a moderate amount of uterine pain, and in
one instance this was repeated at six successive menstrual
periods. Between the menstrual periods these patients

suffered moderately from nausea and other uterine symptoms, the womb was painful when examined, enlarged, and admitted the uterine sound to the depth of four inches in one case. Rest and cooling injections, combined with small doses of ergot and strychnia, caused the purulent discharge to be replaced by the usual menstrual flow. I refer to chronic metritis, cases described by my friend Dr. Matthews Duncan as uterine leucorrhœa of old women. I am justified in doing so, by the discharge being often muco-purulent, and in the only case in which Dr. M. Duncan had an opportunity of examining, he found "the walls of the uterus abnormally thin and soft, and the mucous membrane of the uterine cavity had an irregular and almost ragged surface, the depressions being apparently seats of ulceration." The explanation of such cases is to be found in the fact that in rare instances, the unhealthy lining membrane of the womb continues to secrete, while the process of atrophy progressing in a healthy cervix contracts the cervical canal. When a certain amount of fluid has accumulated the patient feels a girdle of pain, and other uterine symptoms, which are dispelled by the spontaneous evacuation of the fluid, or by its surgical removal, for it is obviously the surgeon's duty to cut short the patient's sufferings by sounding the womb, and it may be necessary in rare cases to keep the cervix distended by sponge tents, until the walls of the uterus have contracted sufficiently to prevent a fresh accumulation. Dr. M. Duncan must have met with this disease, more frequently than I have, and under an unusually aggravated form, for I have seldom seen it, and I believe he exaggerates the danger of such cases, judging at least from a passage of a letter in which he tells me "that if not cured, these cases end in malignant ulceration in the uterus," neither has Dr. West met with the cases described by Dr. Duncan more frequently than I have. His treatment principally consists in the injection into the womb of a solution of nitrate of silver, through a hollow probe, or a double current syringe.

From this state, there is but one step to the obliteration of the cervical canal, which frequently occurs after the méno-pause, without the slightest impairment of health. Should, however, the obliteration of the cervix coincide with an ulcerated state of the uterine mucous membrane, its secre-tions will accumulate, the womb will swell, and then decrease, to enlarge again, as already related, after a few months, with more or less pelvic pain, nausea, and disturbance of the general health. In three cases of this description I did not feel justified in re-establishing the permeability of the cer-vical canal, and under the influence of rest, purgatives, liniments to the hypogastric region, the fluid collected in the womb was absorbed, and the patients have passed many years without presenting signs of any similar recurrence. Dr. Bennet informs me that he never adopted any other plan of treatment in such cases, and the extent to which uterine secretion may be absorbed is shown in a patient of Scanzoni, aged fifty-seven, whose uterus rose two inches above the pubis, to subside to the level of the pubis, after the applica-tion of nitrate of silver to the cervix, and the use of astringent vaginal injections. It is clear from what pre-cedes, that whenever we have to apply strong caustics to the mouth of the womb, after cessation, it is still incumbent on us to maintain the permeability of the cervical canal.

Many of these cases had given rise to no disturbance of health, and had only been revealed on opening the body to discover some other disease ; they will remind the reader of another case, related by Dr. Jallon, *Thèse* 459, An XIII. A woman, aged fifty-two, two months after cessation, had difficulty in passing water and a swelling of the womb, anasarca, and atrophy, followed by death. The womb was found distended, its walls as thin as paper, and containing a yellow curd-like fluid. Duparque has also recorded a fatal termination, in a case of this kind which occurred ten years after cessation. The distended uterine walls burst in a paroxysm of pain, causing death on the fol-lowing day. A large quantity of blood was found in the peritoneum ; the neck was cartilaginous, and the cervix obliterated.

The previous cases enable us to understand the transformation of the womb into a calcareous shell. In nine out of the eighteen cases collected by the elder Louis, cessation had taken place, and the age of the others was not mentioned. The 14th case in Louis' *Mémoire* relates to a woman who was subject to violent hysterical fits at forty, and their subsidence coincided with the formation of a hard tumour behind the pubis, which gave rise to no untoward symptoms. After cessation, she was troubled, for twenty years, with hæmorrhoids, which sometimes bled, and she died at last of consumption. The womb had become a calcareous sac, containing an inodorous, caseous looking fluid ; its shell was four lines thick, and so hard that a hammer was required to break it.

UTERINE DEVIATIONS.—At the change of life or after cessation patients seldom seek advice for uterine deviations, but they are still as frequent as when most complained of. Dr. Saussier de Troyes says, that out of 102 cases of uterine deviations, he has met with eleven women, from forty to fifty, suffering from this disease. I have detected many well-marked cases of anteversion and retroversion at the change of life and after cessation, which did not in the least inconvenience the patients ; and, on the other hand, I have forty patients who, for many years before the ménopause, suffered much, from the displaced womb being diseased, and who are now leading a very active life, although the uterine displacement is as great as ever it was. The cessation of menstruation has extinguished the ever-recurring ovarian irritation and uterine congestion, which were the real causes of their sufferings, and not the displacements or malformations that still persist after the recovery of health. Such facts are calculated to convince uterine pathologists that it would be wrong to consider the displacement as the *chief* indication of the treatment of uterine deviations, which should be governed by the nature of the morbid condition causing or complicating the deviation. This point of practice, and particularly the injudiciousness of treating uterine displacements by intra-uterine pessaries, has been definitively settled by the discussion that took place some years ago in the Imperial Academy of Medicine. I say definitively, for while the discussion was proceeding, Depaul, the reporter of the Commission, asked Sir James Simpson for cases to illustrate his treatment, but could only obtain instruments. Sir J. Simpson never appealed against the verdict then given against him, never alluded to it, nor to the many deaths caused by the use of the intra-uterine pessary, when his works were collected, under his supervision, by two of his best pupils ; he allowed judgment to go by default, and the profession has confirmed the verdict of the Imperial Academy. Although rarely occurring at the change of life, and after the ménopause, I have seen procidentia uteri occur for the first time at the age of forty-nine, in a single lady living in affluence, which could only be ascribed to forcing pains, frequent, though by no means severe, occurring during the dodging time. A dispensary patient, aged sixty-five, suffered as much from prolapsed womb after, as before cessation, which occurred at forty-eight ; she consulted me for flooding, the consequence of a contusion of the prolapsed womb on being knocked down in Holborn. A Mrs. M. has had several children and miscarriages, but never uterine prolapsus till the sixty-ninth year of her age, when it was brought on by a severe attack of diarrhœa. The very rare cases of prolapsus in old age are explained by the fact insisted on by Kiwisch, that although the womb has become atrophied, it is less powerfully supported by a weaker and shorter vagina, and is no longer padded by the fat, which, in youth, abundantly lines the pudenda. This state of the parts likewise accounts for the occasional coincidence of prolapsus of both the womb and the rectum in old women.

Ovaries.—Before passing in review the heteromorphous growths that become so frequent after the ménopause, I will say a few words on the ovaries considered as the seat of congestion and inflammation. I am not aware of there being any case reported of abscess of the ovaries, at the change of life, neither have I noticed subacute inflammation of the same organs. It is difficult to separate the influence of the ovaries from that of the womb in the production of the sexual diseases of women, at the change of life. Nevertheless, as we positively know that the ovaries rule supreme over menstruation, and that they cause many diseases of women during the period of woman's greatest reproductive energy, it is fair to suppose that they aggravate, and delay the cure of the most common uterine diseases, when ovarian irritation is no longer relieved by an habitual menstrual flow. There is no fear of erring in partly attributing to ovarian agency the unexpected reawakening of passion, and the strange sufferings that I have described as ovario-uterine neuralgia. As for the effects of the atrophy of the ovaries, which gradually follows the ménopause, they have been studied in the present chapter, and in that on the physiology of women at the change of life.

UTERINE POLYPI.—They obscure the limitations of the menstrual period, retard cessation, complicate its diagnosis, and are as frequent at the change of life, as in previous periods. This will be shown by the following table, which embodies Dupuytren's experience, and likewise proves the little frequency of polypi after cessation. Polypus of the womb began in—

1 woman from the age of 15 to 20		
10	,,	,, 20 ,, 29
19	,,	,, 30 ,, 39
23	,,	,, 40 ,, 49
3	,,	,, 50 ,, 59
1	,,	,, 60 and above.

57

UTERINE FIBROUS TUMOURS.—Braun, Chiari, Malgaigne,

s

Mr. Paget, and Dr. West, establish the fact that between forty and fifty, uterine fibrous tumours are generally fatal; and I believe that most of these polypi and fibrous tumours originated from thirty to forty, though they may have only become apparent at the change of life. Fibrous tumours retard cessation, and render the occurrence of flooding more frequent, which is not surprising, considering how profuse the menstrual flow becomes so soon as a fibrous body is developed in the womb. I have a patient, now fifty-four, in whom a large fibrous tumour began at thirty-three; and ever since, the menstrual periods, though continuing regular, are extremely painful, and she passes a large quantity of blood, looking like treacle. After cessation the tumour may diminish, provided the change of life has not given it an increased impetus.

S. A., tall, thin, with dark hair, aged fifty-one. The menstrual flow came at twelve, and continued regular but scanty. She married at twenty-one, but never conceived. The menstrual flow remained regular until forty-three, when a fright brought on flooding; the abdomen then swelled, and she was thought pregnant. Flooding continued more or less for eight years, but it never interfered with her appetite, and only lately with her strength. The os uteri is enlarged, but in a healthy condition, and a large, flattened, fibrous tumour is easily detected lying across the abdomen. It may be well to mention that Dr. Bernutz has seen two cases of hematocele at the change of life, which were mistaken for fibrous tumours.

Those who have passed some time at the Salpêtrière know that frequently the fibrous bodies of the womb are found covered with cretaceous deposits, and that these bodies are sometimes replaced by calcareous concretions, which can be enucleated from the walls of the womb. I have twice seen a marked diminution of an ordinary fibrous tumour to coincide with the internal and external exhibition of iodine; syrup of iodide of iron and iodide of potassium, in drachm doses, being given twice a day, while the iodide of potassium ointment was used externally.

CANCER OF THE WOMB.—From forty to fifty is the "cancerous age" of woman's life, for the Registrar-General's Report shows, that previous to thirty years of age cancer is uncommon, both sexes being then equally liable to it; that between thirty and sixty cancer is very common, chiefly among women, and that the relative number of deaths from it, in women and men is—

From 30 to 40 as 19 to 6
„ 40 „ 50 „ 51 „ 6
„ 50 „ 60 „ 5 „ 1

Out of 1200 cases extracted by Dr. Walshe from the Registrar-General's returns, 321 referred to males, 879 to females, and fifty only out of the 1200 occurred before thirty years of age. Leroy d'Etiolles found, that out of 2781 cases of cancer, collected from French authors, the disease occurred 1227 times in persons above forty years of age, and 1061 times in those who had passed sixty. The womb was affected with cancer in 30 per cent. of these cases, and the breast in 24 per cent.

The statistics of Bayle, Boivin, and Lever, confirm the general belief that cancer of the womb is most frequent between forty and fifty; and Dr. West has lately shown that out of 426 cases recently collected by Lebert, Kiwisch, Scanzoni, Chiari, and himself, 178 occurred from forty to fifty, and 122 from thirty to forty. The same fact results from the following table, published by Tanchou, *Gaz. des Hôp.*, 1838.

TABLE XXIX.

Age of 2568 Women dying from Disease of the Sexual Organs in the Département de la Seine, from 1830 to 1835.

Age.	Disease of Sexual Organ.	Cancer.	Remarks.
Before 20	25		{ Maximum of inflammatory affections.
From 20 to 30	442	86	
„ 30 „ 40	279	212	{ Maximum of cancer.
„ 40 „ 50	137	402	
„ 50 „ 60	70	363	
„ 60 „ 70	60	242	
„ 70 „ 80	42	147	
„ 80 „ 90	13	58	

The influence of cessation in producing cancer of the womb is to me evident. Sometimes it coincides with the first manifestation of uterine cancer, as in six out of Lebert's eighteen cases, and in twenty-seven out of fifty-three of my own. Cessation often gives great activity to cancer of the womb that had previously remained undetected, and more particularly then, does cancer progress insidiously and remain undiscovered until its work of destruction is nearly done. At the same time it is fair to admit with Duparque and Saucerotte, that however frequent cancer may be at the change of life, it often originated before; but if the average duration of uterine cancer be about sixteen months, as stated by Lebert and Dr. West, and if, as I have shown, the average duration of the dodging time is three years, then the coincidence between the change of life and the development of cancer becomes obvious. Sir C. M. Clarke considered the corroding ulcer of the womb peculiar to the change of life. With regard to treatment, I shall merely state, that it may be better to narcotize a patient by means of the hypodermic injection of a solution of morphia than to leave her a prey to agonizing pains. When the womb, rectum, and bladder, are united into a foul cloaca, cold water irrigations promote cleanliness, assuage pain, and sometimes induce sleep.

UTERINE HYDATIDS occur so rarely from forty to fifty that I have never observed them; but Dr. Ashley thinks the complaint more dangerous at the change of life.

FATTY DEGENERATION OF THE WOMB.—Two cases have been lately recorded in which the muscular tissue of the womb was in a state of fatty degeneration, at fifty and fifty-three years of age.

OVARIAN TUMOURS.—Like uterine polypi and fibrous tumours, ovarian cysts generally originate in the period of full sexual activity. Thus, Lebert found that out of fifty-nine chronic ovarian tumours,

15 occurred from 20 to 30 years.
12 ,, 30 ,, 40 ,,
13 ,, 40 ,, 50 ,,

From the Fifth Report of the Registrar-General, it appears that out of 100,000 deaths from all causes, there were 1205 from ovarian dropsy, of which 362 occurred from forty to fifty, and out of forty-four deaths, occurring in London from this disease, in 1848, seventeen took place between forty and fifty; thus confirming the received notion that ovarian tumours are most fatal at the change of life, even if they have originated previously. Their march is often more rapid than that of uterine fibrous growths, and more speedily followed by a fatal termination: but, as far as the menstrual flow is concerned, ovarian are far less detrimental than fibrous tumours; for ovarian tumours, when developed many years before the average date of cessation, oftener diminish than augment the menstrual flow; which, in other cases, proceeds uninterruptedly, and is sometimes altogether absent. It seems to me that cessation is rather brought on earlier, than retarded by the growth of ovarian tumours, as I have stated in the chapter on the physiology of the change of life; at all events, they do not cause flooding.

Cases of galloping ovarian tumours generally occur between twenty and forty; and when an ovarian tumour has remained quiescent for some time before the change of life, this will either give it increased activity, or make it stationary. There are many instances on record of women having lived to an advanced age without their comforts being much interfered with by an ovarian tumour; but if it be voluminous or on the increase, it is better the patient should run the risk of ovariotomy.

Taking into consideration that the internal exhibition of medicines is almost useless to check the progress of ovarian tumours, that tapping is only the beginning of the end, and that other surgical measures are of little or no avail, except in very exceptional cases, a pathologist cannot find words too strong to express his admiration for the American surgeons who originated ovariotomy, and for the British surgeons by whom it has been perfected. Mr. Spencer Wells, who must be mentioned as first amongst British ovariotomists, estimates the mortality of his first 300 cases to be 28·33. To ascertain how far this mortality was modified by the patients' age, I have constructed the following table out of 292 of his cases in which the operation had been completed :—

TABLE XXX.
Mortality of Ovariotomy.

Rates of Mortality.	Averages.	Cases.
General average of mortality at all ages .	28·33	300
Average mortality in women under 40 .	31·64	158
Average mortality from 40 to 62 . . .	28·35	134
Average mortality from 40 to 50 inclusive	28·39	81
Average mortality from 50 to 62 . . .	30·76	52

This table makes it clear that the change of life does not raise the mortality of ovariotomy, that it is attended with greater danger, when performed after fifty, and is still more fatal in women under forty years of age. The practitioner will, however, bear in mind that when once ovariotomy has become desirable, it is not wise to defer it long.

MAMMARY IRRITATION AND SWELLING.—This is only noted in fourteen out of my 500 cases, though no doubt it was much more frequent. The breasts are swollen and painful, the nipples sore, and sometimes distil a milky or a glutinous fluid. Friction, pressure, or moving the arm, increases the sufferings, which may be appeased by constitutional treatment, by an atropia liniment, and by wearing cotton-wool next the skin. Spontaneous ecchymosis sometimes marks the breasts at this period, and Dr. Semple has published a case in which a bloody discharge from the nipples continued, every month, for five years after cessation. I attend a lady who for more than a year after the change of life had, every three weeks, a painless exudation of a considerable quantity of red serum, lasting for several days.

Velpeau thought neuralgia of the breast, as well as its neuromatic tumours, were most frequent at the change of life. A liniment or an ointment containing glycerine and sulphate of atropia is the best application in such cases.

NON-MALIGNANT MAMMARY TUMOURS.—Velpeau attributed these to diseased menstruation at puberty and to the change of life, or to pregnancy, and diseases of the womb; and says truly that, after cessation, they cease to increase and often diminish, so their extirpation should not be determined upon too soon. Gardanne relates two cases in which the mammary tumour appeared at cessation, and was cured by leeches. My two cases yielded in a few weeks to leeches and applications of iodide of lead ointment. Mr. Erichsen finds unilocular cysts of the breast to be most frequent from forty-five to fifty.

Velpeau united in the following table 281 cases of hypertrophy, cysts, nodosities, and adenoid tumours of the breast, and found that they were said to have originated at the following ages :—

Up to 30	76
30 ,, 40	64
40 ,, 50	80
50 ,, 60	19
60 ,, 80	31
Age not noted	11
	281

Cancer.—Galen observed, that cancer of the breast was frequent at the change of life, and Mr. Paget has lately shown that it occurs most frequently from forty-five to fifty years of age.

TUMOUR OF THE MEATUS URINARIUS. — I have met with three cases of erectile tumour of the meatus urinarius, a disease thought by Sir C. M. Clarke to occur most frequently in the earlier part of woman's life; but I am confirmed in the contrary opinion by Mr. B. Norman, who has observed fifteen cases, eleven of which were in women during or after the change of life. Four of my five cases came under treatment after the ménopause, and in a case that I attended with Dr. Bennet, the tumour, successfully treated by him several years before the change, returned after the ménopause, and was so very hard that we thought it preferable to cut it away, which was skilfully done by my friend. At the time of the operation there was very little bleeding; and when, on the following morning, it became necessary to draw off the urine, I found it largely mixed with blood, and as this continued to be the case during the following days, I came to the conclusion that the blood had been oozing from the wounded urethra into the bladder. At all events, one application of the undiluted perchloride of iron prevented any subsequent mixture of blood with the urine. Diseases of the bladder and urethra so frequently afflict men from the fortieth to the sixtieth year, and women of the same age are so free from such diseases, that this may be viewed as compensating favourably for the many other diseases to which women are specially liable.

DISEASES OF THE KIDNEYS.—The forty-nine cases in which women were, for a few months, habitually subject to an increased amount of the urinary salts, are rather to be considered as instances of a critical evacuation from the blood than as proofs of diseased action, and the same applies to *Hæmaturia*—a rare disease—which has been observed by Chouffe, Menville, and Dusourd. My cases yielded to bleeding and mild treatment. The eight cases of pain in, and difficulty of, passing water, and those of incontinence of urine, soon yielded to diluents, baths, and mild measures.

The following case of inflammation of the pelvis and calices of the kidneys is instructive :—

CASE 46.—The wife of a physician, who was born and had passed the first fifteen years of her life in India, enjoyed excellent health until the change of life, which unfortunately coincided with the sudden loss of a dear sister and an unusually cold spring. The result of which was a sluggish liver, dyspepsia, pains in the bladder, dysuria, and abundant deposits in the urine. As this condition did not abate, the husband consulted an eminent surgeon, who tested the urine by nitric acid, and said the patient was suffering from albuminuria. This diagnosis was unfortunately told her, for she had dabbled in medicine, and knew that albuminuria meant an ugly future, so she gave herself up to despair, made her will, settled the distribution of her trinkets, and kept her husband without sleep for two nights, which was decidedly useful, as it suggested to him the advisability of consulting Dr. Owen Rees. He also examined the urine, not neglecting to use the microscope, and he found there was pus in the urine, and he supposed it to be caused by a renal calculus. Two months of careful dieting, Vals water, and rest, removed pus from the urine and all symptoms of disease, and as there never was either persistent pain or paroxysms of pain in the region of the kidneys; as there was no bloody urine, and no calculus was ever passed—may be there was none—but merely inflammation of the pelvis and calices of the kidneys. Towards the following autumn the patient had a relapse, so she went to the south of France and derived so much benefit from passing the winter in a milder climate that she has continued to do so.

CHAPTER X.

DISEASES OF THE GASTRO-INTESTINAL ORGANS.

TABLE XXXI.

*Liability to Diseases of the Gastro-intestinal Organs in
500 Women.*

Monthly toothache and swollen gums	1
„ Water-brash	5
Hæmatemesis	4
Repeated vomiting of mucus	31
Jaundice.	6
Long-continued biliousness, induced or increased	55
Dyspepsia	37
Monthly dyspepsia	1
Obstinate constipation	23
Frequent diarrhœa	45
Monthly diarrhœa	5
Entorrhagia	20
Monthly blood in motions for six months . . .	2
Pus in motions	3
Inflammation of the rectum	3
Blind piles	62
Bleeding piles	24
Piles bleeding every month	1
Swelled abdomen	26
Total	354

This table means that 354 kinds of gastro-intestinal suffering were noted in 500 women at the change of life, and
many of the 354 presented several kinds of gastro-intestinal

disturbance.   The practical bearing of this table is the
great liability to biliousness, dyspepsia, diarrhœa, and piles.
Later in life, the tendency to jaundice, biliousness, and dyspepsia diminishes and appetite and digestion improve; there
is also less tendency to lose blood by the bowels; piles
cease to bleed, and generally subside in advanced life.

BILIARY AND DYSPEPTIC AFFECTIONS.—The liability to
derangements of the biliary apparatus at this period has
been noted by Burns, Gendrin, Meissner, Otterburg, and
Aran; by Sir C. Mansfield Clarke, Sir J. Simpson, Dr.
Evory Kennedy, and Dr. West.   Hepatitis and jaundice
have been observed by Gardanne, and Sir H. Halford mentions having seen abscess of the liver at cessation.

The late Dr. B. Lane's observations on the coincidence
of the derangement of the biliary secretions at the change
of life, are true to nature :—" Nothing can be more common
than to find severe biliary derangement occurring at or
about the period of menstrual cessation; and, looking at
the great physiological change which then takes place in
connexion with hepatic development, it is naturally to be
expected.   A woman will complain of being bilious—there
may be a bitter, oily taste in the mouth, a burning in the
throat, frontal headache, nausea, and even vomiting, the
urine high-coloured, the bile abounding in the alvine dejec-
tions, and perhaps causing heat and a stinging sensation in
the rectum, the tongue furred, a biliary tinge pervading the
cutaneous surface."

The worst cases of disease at the change of life for which
I am consulted have been ascribed to a torpid condition of
the liver by other medical men; and though the preceding
table shows clearly how liable women are to sickness, dyspepsia, and biliousness at this epoch, it cannot give an idea
of the obstinacy of the biliary symptoms in many of these
cases.   Thus, P. K. was a strongly-built woman, of a sanguine
temperament, in whom the menstrual flow had been irregular
for the last eight months; till then she had enjoyed good
health, but since, in spite of purgatives, alteratives, and

tonics, I found it difficult to set right the gastro-intestinal
functions, or to improve the appearance of the tongue, which
was permanently coated with a yellow fur.

CASE 47.—*Jaundice.*—Mrs. W. is a tall, stout lady, with
dark hair, forty-nine years of age, and the mother of thirteen
children.   The menstrual flow appeared at twelve, and
continued regular until it ceased suddenly at forty-seven.
Six weeks afterwards the skin became yellow, and she
suffered considerably in the region of the liver.   She was
soon cured, but every three or four months there was a
return of jaundice.   This lady is habitually low spirited,
and so drowsy that she can neither read nor do fancy work
during the day, and is restless at night.   She has often
flushes and chills, but no perspirations.   After diminishing
the size of the engorged liver, and curing the actual attack
of jaundice by blue pill and alkalies, the question was, how
to prevent that congestion of the liver which had returned
with a certain kind of periodicity, since the sudden cessation
of menstruation.   I gave blue pill and alkalies for a week
every month, and I established a free determination of blood
to the skin, so that it might diminish the congestion of the
internal organs, by warm baths, prolonged for two hours
every other day, and daily scruple doses of the comp.
sulphur powder.   I advised 2 oz. of blood to be taken from
the arm every third or fourth month, as might be otherwise
indicated.   Gentle perspirations soon showed that the blood
currents were steadily setting in a right direction, health
became good, and for the last two years there had been no
return of jaundice.   The same plan of treatment succeeded
equally well in the following case.

CASE 48.—*Repeated jaundice at cessation.*—Mrs. H., a
robust lady, with a bilious look, consulted me in June, 1854,
being then fifty-five.   The menstrual flow came at fifteen,
and continued regular until she was forty-eight, when it
ceased suddenly.   She was then troubled with piles, which
bled freely every six or eight weeks.   This went on for six
months.   When the piles ceased bleeding, her skin became

yellow, and the liver was found extending several inches below the edge of the false ribs. She has had several slight fits of jaundice, and her motions are often black, offensive, and " burn" the passage. It should be observed that both this and the patient in the previous case, had never been particularly bilious before cessation. The cause of long-continued biliousness at the change of life, is the congestion of the portal system, no longer relieved by the menstrual flow, so that the liver is overworked. Thus I have seen jaundice occur in young women from the sudden suppression of the menstrual flow, and have sometimes noticed a persistent loss of appetite for months, notwithstanding the administration of alteratives and tonics. It must also be noted that when uterine diseases occur at the change of life, they are more frequently attended by biliary derangement, and by the secretion of an enormous quantity of bile.

TREATMENT.—The frequency and tenacity of biliary and dyspeptic affections at this period, explain why purgatives have been considered the principal medicines required; but an occasional purgative will have little effect, a systematic plan of treatment must be resorted to, and calomel or blue pill associated with the comp. ext. of coloc., or with the soap and aloes pill, is well borne and often required. Since my last edition, there has been an attempt to prove that mercury does not promote the flow of bile, but I submit that experiments made on healthy dogs, do not apply to diseased men and women. It is a matter of daily experiment to find, that while various purgatives repeatedly taken, bring away no bile, one dose of calomel will cause yellow, burning bile to pass in the motions, so I consider calomel and blue pill to be indispensable for the treatment of diseases of the liver at the change of life. The only tonics that should be given while the liver is thus engorged, are the diluted mineral acids, taken alone, or in some slightly bitter infusion, three times a day, before meals, and immediately after them fifteen drops of liq. potassæ, in a wineglass of water, and one or two effervescing draughts, with citric acid and carb. of soda and potash, should be also taken during the day. They may be flavoured with syrup of ginger, or syrup of orange peel, or ten to fifteen drops of chloric ether. Instead of porter and beer Seltzer water with light wine should be advised. Prolonged tepid baths are useful, and an occasional small bleeding may favour the action of the remedies. The late Dr. Wright attributed great efficacy to hydrochlorate of ammonia, which he gave in a mixture with hydrochloric acid. I have seen cases which resisted all treatment cured by a course of mineral waters at Vichy, or at Homburg, or Kissingen ; the action of the waters being admirably assisted by early hours, plain diet, and the complete change of all the associations of life.

VOMITING.—This is not uncommon during the change of life, sometimes accompanying the menstrual flow, and replacing it at others. This symptom generally yields to morphia, a twelfth or a sixth of a grain of the acetate being given in an effervescing draught, and repeated until sleep be induced. A patient, who for years had suffered from vomiting at the menstrual periods, is always relieved by less than half a grain of the acetate, while two grains taken in the twenty-four hours, will not always quell the vomiting of another patient. One patient had no other symptom than water-brash, which appeared at cessation, and recurred every month for eight months. Hufeland and Menville speak of the liability of women to hæmatemesis, which occurred in four of my cases.

DIARRHŒA.—This has been noted as occurring in an habitual manner, at this period, by Gendrin, B. de Boismont, and Chambon. It may constitute the only symptom of cessation, and should always be considered a critical discharge, and Portal saw its suppression by active remedies bring on anasarca. The advantages and disadvantages of diarrhœa at the change of life are well shown in the following cases :—

CASE 49.—*Chronic diarrhœa.*—A tall, stout, and florid lady, aged fifty-seven, showed an hæmorrhagic tendency by the abundance of the menstrual flow, the liability to dysentery, and by her numerous confinements being always accompanied by a flooding. When fifty, the lady met with severe family misfortunes, flooding ensued, the menstrual flow ceased to be regular, occurring, even now, as a mere show, after nervous excitement. Soon after fifty, she suffered from erysipelas in the head; and as a result of this, of cessation, and of mental shock, the lady remained long in a very nervous state. Ever since the ménopause, for the last four years, this patient has had three or four motions every day without pain or any debilitating effects, the appetite and digestion continued excellent, and the increased mucous flow from the intestines sufficiently relieved the constitution, as the health remained good, notwithstanding the absence of perspirations until lately.

CASE 50.—*Diarrhœa aggravated by cessation.*—Sarah C. is a thin, nervous, chlorotic-looking single woman, of forty-seven. The menstrual flow appeared at fourteen, and continued regular until six months back, when it became very irregular. Flushes, perspirations, nervousness, and irritability were then troublesome, and twice during this period she has had to leave her place on account of diarrhœa. She states that, ever since the menstrual flow first appeared, she has always had from three to six motions a day, but was always most troubled at the menstrual periods. Until lately, however, this state of the bowels never interfered with her health or appetite, but now, six motions occurring daily, completely incapacitated her for work, although they were unattended by pain. I moderated the diarrhœa by blue pill and chalk mixture, and then prescribed my usual remedies with success.

Dr. Cockerton, of Montgomery, consulted me respecting a lady, aged fifty-three, who, during the dodging time, suffered from dysentery, then flooding, and, on its subsidence, from great relaxation of the bowels, with racking pains in the lower part of the abdomen, which occurred regularly every morning, and were evidently connected with uterine disease.

Chambon also has seen the constitution thoroughly shattered by diarrhœa at the change of life. I have met with women who, in advanced age, had several semi-fluid motions in the course of the day, without suffering in the least, and Dr. Day has met with similar cases.

TREATMENT.—I seek to restrain the diarrhœa by diet and abstinence, but if that be insufficient, chalk mixture should be given, and after that, small doses of Dover's powder with blue pill made up into pills. Extra clothing and exercise, a warm bath, and, if the patient be plethoric, a small bleeding may be useful. The following cases will show the plan to be adopted in cases of Entorrhagia :—

CASE 51.—*Periodical entorrhagia.*—Miss M., thin, nervous, and chlorotic, consulted me in 1853, being then forty-eight. The menstrual flow appeared at fifteen, and continued regular until about three years ago. For a year the flow was sometimes more, at others less, and it suddenly ceased two years ago, after a succession of frights, which thoroughly shook her nervous system. Her health, good until cessation, has been bad ever since. She is often troubled with cough, leucorrhœa, and loss of power in the lower limbs. She has become unusually nervous, feels, at times, as if she were going into fits, talks of strange sensations in the head, as " if she had taken something to make her silly." This nervous state is peculiarly marked every three months, when she menstruates by the bowels for about six days. Several times a day she feels great pain at the lower part of the abdomen, there is then anal tenesmus, and she passes a small quantity of blood and mucus. This has occurred for the last two years, and during all this time there have been no flushes nor perspirations. In this case, nature actually did for two years what I have sometimes recommended—to take a small quantity of blood from the system. I gave her the comp. camph. mixture before, the carbonate of soda after meals, 3 grs. of blue pill, with 2 of ext. of hyosc., on alternate nights, and a warm bath for an hour, every other day. This was continued for about a month with great amendment of the nervous symptoms, and a week before the time for the passing of blood by the bowels, I had 4 oz. of blood taken from the arm, which prevented the hæmorrhage. I then advised the mixture and carb. of soda to be taken as before, and so much of sulphur and borax as would ensure a comfortable action of the bowels. In a few weeks gentle perspirations had become habitual, and afforded great relief. This patient took the syrup of citrate of quinia and iron, in 60 drop doses, twice a day for two months, and became comparatively strong. I had again 4 oz. of blood taken away before the hæmorrhage was due, and lately, when I saw the patient, she was in good health.

Case 52.—*Repeated entorrhagia and hæmoptysis.*—Miss G., a tall, delicate-looking lady, consulted me in 1852. The menstrual flow appeared at sixteen, and it had been no source of trouble until two years ago, when it became irregular, sometimes missing for two or three months, sometimes being scanty, at others so abundant as to amount to a flooding. After shivering and griping pains, this patient has sometimes passed considerable quantities of blood and mucus, and once she brought up a large quantity of blood from the lungs. Since cessation took place, the passing of blood from the bowels has not been more frequent, and notwithstanding continuous flushes and perspirations, she is very nervous, is troubled by sensations at the pit of the stomach, as if some one were sitting there, has choking sensations never before experienced, is very low-spirited, has crying fits, and though sleeping pretty well at night, could do so all day. What, however, the patient most complains of is, frequent recurrence of pain at the lower part of the abdomen, and for the last few nights the legs, the pudenda, and the seat, are much swollen. This led me to anticipate a recurrence of the entorrhagia, so I ordered a hot linseed poultice to the lower part of the abdomen at night, and an ounce of castor oil to be taken in the morning, during which day she passed " blood and corruption," and continued to do so with a great deal of griping for five days, after which all her abdominal suffering vanished. All her life, this patient had never less than three

T

motions a day; the menstrual periods always ending with
five or six motions a day; this being the habit of the body, it
is easy to understand that for the last year, ever since cessation, the system should seek relief by the intestinal surface.
With a view of remedying this state, I ordered the co. camph.
mixture before, and the carb. of soda after meals, 3 grs. of
blue pill, 1 of ext. of gentian, and 1 of ext. of hyoscyamus,
every other night; 10 grs. of Dover's powder every other
night; a belladonna plaster to be applied every week to the
pit of the stomach, and warm baths to be taken for an hour
on the alternate nights. This treatment was continued with
considerable advantage for a month, when the baths and
plasters were omitted, and the pills were only taken occasionally. Two months after I first saw the patient, I had
4 oz. of blood taken from the arm, and this was repeated
three months later. I last saw her about a year ago, and
she was well; neither the hæmorrhage nor the menstrual
flow had returned.

CONSTIPATION.—This was observed in twenty-three cases,
and yielded to the continued use of mild but varied purgatives. In some cases the patients suffered severely from
colics, as was the case with Mrs. T., aged forty-seven;
the menstrual flow coming at fifteen, had left gradually at
forty-five. Ever since cessation she has been constipated,
instead of relaxed, as before; and during the last year she
has been troubled every two or three weeks with colic,
flatulence, and sometimes vomiting. These symptoms
coming on every night with greater severity, I prescribed
pills containing 1 grain of ext. of opium and 3 of c. ext. of
colocynth. Two pills nightly were at first required to lull the
pain, and then one sufficed. This plan of treatment was
effectual, the bowels being afterwards kept moderately open
by the habitual use of sulphur.

HÆMORRHOIDS.—Gardanne has mentioned the case of a
lady in whom cessation occurred at forty-eight; and, until
seventy-five, when he attended her, she lost monthly a large
quantity of blood from piles, and suffered from signs of

plethora until the evacuation took place, and Stahl has seen similar cases. Menville gives one, in which, for ten years after cessation, hæmorrhoids occurred monthly. Gendrin and B. de Boismont have likewise noticed continuous or intermittent hæmorrhoidal fluxes at cessation. The table which heads this chapter shows how frequently I have met with piles. The hæmorrhoidal discharges, the diarrhœa, the biliary affections of the change of life, all arise from one common cause—the unrelieved plethora of the portal system; and when we bear in mind that the organs of reproduction, as well as the intestines, are principally innervated by ganglionic nerves, that the spinal nerves of the womb and the intestines arise from the same part of the spinal cord, that the veins of the uterus communicate mediately with the portal system, and that the last portions of both canals are contiguous, it is not surprising that, when the uterine discharge is arrested, the nervous energy and the sanguineous current which used thereby to find vent, should flow to the intestinal surface.

TREATMENT.—Although critical, the hæmorrhoidal flow is so disagreeable, that it should not be permitted to endure longer than can be helped, and in most cases it can easily be remedied by following my general rules of treatment. The occasional removal of 3 oz. of blood from the arm is clearly indicated, so is the compound sulphur powder, so useful in hæmorrhoidal affections, and rectal sedative injections to allay the local irritation, which continually draws the blood within its morbid circle of attraction. Cooling and astringent lotions and ointments to the hæmorrhoidal tumours are most useful.

INFLAMMATION OF THE RECTUM.—Without being subject to piles, women sometimes feel great heat, weight, and tenesmus in the rectum. It will sometimes cause a difficulty of passing water, and a sensation as if a foreign body were in the vagina. With repose and cooling lotions, these symptoms generally abate; though intractable cases will sometimes occur like the following, which is a

T 2

remarkable instance of severe suffering caused by the change of life.

CASE 53.—*Inflammation of the rectum.*—Charlotte O., the wife of a plumber in good circumstances, is a short, average-sized woman, with brown hair, flushed, puffy, and damp face, downcast looks, seems drowsy, and looks as if she had been drinking; was forty-four. The menstrual flow appeared at eighteen, after pains flying about her; and soon after its appearance she suddenly felt something shoot through the ears, and her neck became stiff. She has never since been able to turn her head freely; and when she tries to do so, she feels something crack. The menstrual flow came regularly, but a fright would at any time cause it to appear. She married at twenty-three, and seven months afterwards she flooded, and lost much blood for two months; since then the catamenia often lasted for three weeks, and were accompanied by a difficulty of passing water. She has consulted many doctors, but no accurate examination was ever made. She never conceived, and has had no serious illness until lately. Although happily married she has become more nervous, more easily excited by the everyday occurrences of life; without ever having had any hysterical attacks. For the last seven years she has been frequently troubled by uneasy sensations "of fulness and rawness" at the pit of the stomach, the weight of the bedclothes annoying her. The menstrual flow has been irregular for the last eight months, coming every two or three months, and twice she has been flooded. Since then, she has been more subject to headaches, and to pain at the nape of the neck. Now she is often light-headed, giddy, forgetful; and so drowsy, that she is obliged to lie down, but only dozes a little. She is more sensitive to all stimuli, starts at the least noise, sometimes gets out of bed to stop the ticking of the clock, and cannot bear to be spoken to. Her nights are bad; and of late, three or four times a week, she is awoke suddenly as if by the ringing of a bell in her ears, then she faints off, and it often takes her husband half an hour to bring her to her senses. Since the menstrual flow stopped a year ago, the epigastric sensations, the flushes, and the sweats were very frequent. There was pain on micturition, and passing stools, and always more or less leucorrhœa, with bearing-down pains. On examination, I found the bladder free from all undue pressure. The womb had the size, form, and appearance of a healthy virgin womb. I therefore considered this to be a case of intense functional derangement, caused by the cessation of the menstrual flow; and ordered her to lose 8 oz. of blood, and to follow up the bleeding by the application of five leeches to each side of the neck; to take the c. camphor mixture before meals, 3 grs. of blue pill, and 2 of ext. of hyoscy. every night; also 10 grs. of Dover's powder. Warm baths were afterwards ordered, but discontinued, because the patient nearly fainted so soon as the water reached her waist. Pediluvia were frequently given, and the nape of the neck and back part of the head were rubbed with camp. liniment, containing 1 drachm of laudanum to the ounce. The patient improved under this treatment, and in the course of a few weeks she had no more nervous fits at night. The pills were then omitted, the Dover's powder only given every other night; the mixture was continued, and she was told to have 3 oz. of blood taken from her every month. I lost sight of this patient until August 26th, 1853, when she was again suffering from many of the symptoms previously described. Three months before, a fright brought on a flooding, which lasted three weeks. No menstrual flow since, but distressing pelvic pains. On examining the womb, I found it healthy, but pressure to the rectum gave intense pain; and she said she had passed a membranous substance from the bowels. The camphor mixture was ordered, a teaspoonful of sulphur and borax every night, a scruple of nitre in a little barley-water twice a day, and a large belladonna plaster to be applied to the pit of the stomach. Sept. 19th.—Tongue furred, very yellow, constipation, bearing-down pains, sensations " as of fire, in the back passage." On making a careful examination, there was no fistula or fissure, the sphincter was hot and contracted, the rectum also felt hot, and the finger gave pain. The mixture was ordered before meals, carbonate of soda afterwards. Blue pill and hyoscyamus every night, to be followed by half an ounce of castor oil in the morning. Three times a day injections were to be made into the rectum, with half linseed-tea and half a strong decoction of poppy-heads; and a little later, with equal quantities of saturnine lotion and decoction of poppy-heads. Various other injections were tried to relieve the distressing symptoms which often returned; but nothing succeeded so well as making a full injection of tepid water, to cleanse the bowels, and then injecting an ounce of the following solution, with 2 oz. of warm water :—

Lotio plumbi acetatis . . . . . ℨiv.
Tinct. hyoscyami . . . . . . ℨiv.

Feb. 17th, 1854.—The menstrual flow had been absent for the previous eighteen months, when, two months ago, she had a pale red vaginal discharge, which returned the following month. She came to the dispensary on account of a relapse of the inflammation of the rectum, which yielded to the measures previously described. In May, 1856, I again saw the patient for a slight relapse; and as she complained of a yellow discharge, I examined, and found the vagina shorter and narrower, and the neck of the womb smaller than on a previous examination. Pressure on the rectum gave great pain, and indicated the seat of the disease. The patient soon got better; but during the five years I have watched this case, I have seldom met with a more distressing example of suffering caused by the change of life. Now, her health is in every respect better; she is forty-eight, and there has been no menstrual flow for more than three years.

I have had several other cases of inflammation of the rectum, in which I have found the daily injection of one ounce of a lotion, containing sulphate of zinc and hydrocyanic acid, was very useful, and at night I made the patients apply to the sphincter, a small quantity of cold cream, to each ounce of which was added one grain of sulphate of atropia.

PERMANENTLY SWELLED ABDOMEN.—This condition, at the change of life, is generally to be accounted for by an increased deposit of fat in the omentum and in the abdominal walls,—a physiological condition treated of at page 59. Sometimes the swelling is principally caused by distressing flatulent distension of the bowels, without diarrhœa or constipation. I have known this to recur every month for several years after cessation, lasting two or three days, and yielding to carminatives, tonics, and such measures as insured the gentle action of the bowels.

CHAPTER XI.

TABLE XXXII.

*Liabilities of the Skin to Disease at the Change of Life in
500 Women.*

Flushes	287
Dry flushes	14
Legs and feet burning, and very painful . . .	2
Hands painfully hot	3
Distressing aching under the finger nails . . .	2
Peeling off of the nails	4
Falling off of all the finger nails	1
Perspirations	201
Monthly ditto	2
Cold ditto	13
Sweats	89
Undetermined cutaneous eruptions	18
Nettle-rash	5
Erysipelas	4
Eczema brought on	3
„ increased	2
Ecchymosis	3
Shingles	1
Herpes circinatus	1
Prurigo	3
Pruritus, or itching without apparent lesion . .	5
Carried forward	663

Brought forward	663
Œdematous, or swelled legs	16
Monthly swelling of legs for three days for a year	1
Œdematous, or puffy face	3
Inflammation of legs, and painful distension of their superficial veins.	3
Varicose veins induced	1
„ „ aggravated	1
Ulcerated leg	7
Boils in seat, and in other parts	3
Abscess in finger	2
„ armpit.	2
„ neck	2
„ groin	1
Total	705

Thus, 500 women divided 705 modes of slight or severe cutaneous suffering at this epoch, many suffering in various ways at the same time. The liability to cutaneous disease passes off in after life; and though flushes and perspirations may occasionally occur, they are no longer a source of annoyance. Prurigo and eczema are, however, sometimes very troublesome to women in advanced age.

These affections may be registered under three heads—flushes, sweats, and cutaneous affections. I shall be brief on the last, but as the flushes and sweats of the change of life have never been sufficiently studied, I shall devote a few lines to their pathological import, referring the reader to p. 63 for further details.

FLUSHES.—This is the popular name, and I adopt it, because it is short and expressive. "Hot blooms," is another expression, which faithfully indicates what really occurs. Flushes are mentioned by nosologists under the name of *ardor volaticus*, or *fugax*. Romberg correctly calls it one of the cutaneous hyperæsthesiæ, and notices its greatest frequency at the change of life. Flushes may be increased to a painful extent by external heat, over-clothing, hot rooms, hot drinks, and over-feeding, by the checking of diarrhœa or of leucorrhœa. Some feel so faint under their influence, that they must have air, or they would swoon. M. C. could drop down with weakness when the flushes came on. Dry heats are morbid flushes; they torment a patient without the subsequent relief of perspiration. The cheeks are not the only parts susceptible of burning sensations, for two patients complained most of the burning of the legs and feet; cold water applications did not much abate the annoyance, but warm did; and at night they slept with their feet out of bed. Three others complained bitterly of similar sufferings in their hands, and of aching under their finger nails. The flushes are sometimes preceded by chilly sensations, and some women tremble with internal cold, and remain, for a long time, habitually cold, notwithstanding the flushing of the face. One patient continued in this state for seven years; she had slight flushes without perspiration, was very nervous, and often fainted. Others feel so cold that they approach the fire, reaction then soon coming on, they fly to the window. In such cases, the perspirations will be cold and clammy, denoting debility, and the congestion of some internal organ. When the flushes are thus anomalous, they are often preceded by ganglionic dysæsthesia; strange sensations, which have been said to resemble " pulses, like a live animal throbbing in the stomach," or "the fluttering of a bird;" sensations which vanish on the appearance of perspiration.

SWEATS.—Perspiration was carried to a morbid extent in seventy-nine out of 500 women, who often complained of heavy perspirations. They were constantly wiping the sweat off the face, their hair was often wet, they were obliged to change their linen twice a day, and although slightly covered, their bedclothes became soaked. Gardanne, Chambon, and B. de Boismont, notice the occurrence of sweats at this period, but the little importance attached to them is shown by the fact, that in an elaborate article, with a host of references, on idiopathic sweating, J. Frank merely quotes Tissot's having observed it at the change of life. It would seem that sweating ought to be more beneficial than perspiration, but a very accurate observer, Sanctorius, remarks —"That perspiration which is beneficial, and most clears the body of superfluous matter, is not what goes off with sweat, but that insensible steam or vapour which in winter exhales to about the quantity of fifty ounces in the space of one natural day." "Sweat is always from some violent cause; and as such—as static experiments demonstrate—it hinders the insensible exhalation of the digested perspirable matter."

The intensity of the force which impels the sweat from the skin is shown, not only by the length of time it may last, but by its often resisting all the attempts made by women to diminish or suppress it. They wear less clothing, take cold drinks, place themselves in draughts, but still the impulse is stronger than their efforts, for they seldom succeed. In general, it is some internal focus of active congestion which checks perspiration, or renders it cold and clammy. When this occurs, great debility is felt, with an increase of epigastric pain in some, in others, pseudo-narcotism; which, as Dusourd remarks, is immediately relieved when the skin perspires. It is very strange that these sweats are sometimes prolonged for months, without weakening women, or preventing their becoming fat, though the fat often seems to be half liquid. Women of a nervous or chlorotic type, have habitually a cold clammy perspiration, like that coming from the blanched skin of those who are vomiting or in syncope; a passive permeability of the skin, determined by loss of nervous power.

TREATMENT.—I have shown the advantages of promoting perspiration to check the serious disturbance of the internal organs, how then should it be restrained so as to inconvenience the patient as little as possible? Blood is the fuel, the interstitial and molecular combustion of which keeps alive the continuous heat of the body. The relative super-abundance of blood is often the cause of the superabundant heat, and therefore of the perspirations, as shown by the case related by J. Frank, of a man who became subject to excessive perspirations after the suppression of an hæmorrhoidal discharge. Diminish the mass of blood by taking three or four ounces from the arm at two or three months' interval, and the sweats will also diminish. The irritability of the nervous system may be relieved by the sedative preparations already recommended. Baths are very useful. They should be tepid, and prolonged for an hour at least. Tissot recommends them highly, and their effects may be increased by adding from one-half a pint, to a pint of camphorated vinegar. The saline matters, which are otherwise removed by perspiration, must be directed to the kidneys, by giving the salines, as already indicated, the acidulated and effervescing drinks, and small doses of nitre and borax. I cannot too strongly recommend mineral acids for those who suffer much from heats and sweating, for they tone the nerves as well as cool the blood. The greatest amount of testimony runs in favour of sulphuric acid, which I have found most effectual. Thus Pereira bears witness to the sedative action of sulphuric acid in distressing tingling of the skin; Hufeland gave Haller's acid elixir, which is mixture of one part of sulphuric acid, in volume with three parts of spirits of wine; other Germans give at the change of life, Mynsicht's elixir, in which sulphuric acid and alcohol are associated with aromatics. I order from 10 to 20 drops of the diluted acid to be taken in a wineglassful of cold water, three times a day, or else the acidum sulphuricum aromaticum of the British Pharmacopœia, but these remedies require time, and will not satisfy a patient whose cheeks burn repeatedly in the course of the day, and who wants something to give immediate relief. Then a lotion of 1 ounce of cherry-laurel to 5 of elder-flower water, may be tried, or water in which camphor has been allowed to float, or vinegar and water, or water containing 1 or 2 ounces of camphorated vinegar to the pint. Chomel advised the treating the hot cheeks by a kind of steam douche, and I have found it useful to do so two or three times a day ; in the same way burning feet

and hands are soothed by soaking them in hot water. Some
patients have derived great benefit from using, with the
powder puff, one of the following powders :—

Flush Powder, No. 1.		Flush Powder, No. 2.	
Carmine. . . ½ gr. or less.		Carmine. . . ½ gr. or less.	
Nitrate of bis-		Camphor . . ʒss.	
muth . . . ʒj.		Oxide of zinc . ʒj.	
Camphor. . . ʒss.		Otto of roses . gtt. j.	
Oil of bitter al-		Starch . . . ʒij.	
monds . . gtt. ij.			
Starch . . . ʒij.			

Some describe with energy their sufferings in winter nights,
when, bathed in perspiration, they are afraid of turning, lest
they should be chilled by the damp cold, and have given me
their warmest thanks for suggesting that a long, thin, flannel
dress be worn over the night-gown, and that they should lie
on a horse-hair or a spring mattress, instead of on a feather
bed. Suppressed perspiration may be recalled by a warm
bath, by letting the patient sit, wrapped in a blanket, on a
common cane chair, under which a lighted spirit-lamp is
placed, and in ten or fifteen minutes the skin will perspire
profusely. Gentle, continuous perspiration may be promoted
by sulphur and sudorifics.

CUTANEOUS AFFECTIONS.—Out of 500 women, forty-one had
some form of cutaneous disease, but whether the proportion
would have been smaller in 500 men, between forty and fifty
years of age, I cannot determine, though I have certainly
found eczema very intractable and relapsing, and prurigo
very troublesome at this epoch, and both complaints more
common than would appear from my statistics. Dr. Ashwell
thinks skin diseases by no means rare at the change of life.
Dusourd and B. de Boismont say that lupus is frequent ; and
the latter observes that herpetic affections, long forgotten,
then reappear. Gendrin, Gardanne, and myself have repeatedly seen erysipelas at this period, and a case is cited by

Tissot of erysipelas of the face occurring fifteen times during the two first years after cessation, less frequently in the two next years, and only once during the fifth year. Mr. Erasmus Wilson, on the other hand, informs me, that he does not consider women more liable to cutaneous diseases at this critical period than at any other, though when such diseases do arise, they are peculiarly obstinate. Prurigo and eczema he deems the most frequent. Mr. Harvey has frequently seen eczematous eruption of the auricle, and behind the ear, begin at the change of life, last for many years, and resist all treatment, until at last it disappeared by a spontaneous effort of nature. This has occurred several times in my own practice. Mr. T. Hunt assures me that, the change of life has often been followed by acne rosacea, lichen in the face or elsewhere, prurigo, and especially prurigo pudendi. He says, that chronic, scaly affections of the skin, such as lepra and psoriasis, which have commenced before this critical period, are unaffected by it, and that cutaneous affections seldom disappear spontaneously at this epoch. A patient of mine had never had the slightest rash before the ménopause, whereas nettle-rash appeared four times in the year which followed cessation; I believe it, therefore, to have been caused by this crisis, as erysipelas was in Tissot's case. M. B., aged fifty, first menstruated between her eighteenth and nineteenth years, with little previous disturbance, and continued regular until twenty, when she married. She had nine children, the last when forty-four. At forty-eight, she had several floodings, but without much increase of pains in the head. The catamenia ceased at forty-nine; this was followed by no disturbance of health, except three months after, by a severe attack of nettle-rash, on the chest and body, which disappeared on the proper medicines being administered; twice, however, it has recurred at irregular periods, and she applied for relief, at the Farringdon Dispensary, for a fourth well-marked attack of the disease on the lower part of the body and the thighs. Alibert observed some cutaneous eruptions appear twice only in life—

once before first menstruation, and again at its cessation. I have also twice known both epochs to be preceded by an abundant eruption of boils, and Gardanne mentions the same occurrence. That form of lupus which attacks the vulva, and was confounded with cancer, until the distinction was clearly made by Huguier, is as frequent at the change of life as at any other period, but is not observed after fifty. Amongst the poor, ulceration of the leg, without being peculiarly intractable, is not an unfrequent complaint at this epoch. Œdematous legs I have frequently noticed, and I have had two cases of varicose veins occurring then for the first time. For the local treatment of prurigo, eczema, and other cutaneous affections of the pudenda, I refer to that of the affections determining nymphomania, p. 238. The practitioner will bear in mind that pathologists are adopting what has been long taught in Vienna and in Paris, that prurigo senilis depends on pediculi, and that the disease is therefore propagated by the patient's clothes. Patients occasionally take strange fancies about insects. A dispensary patient was intensely miserable for many months, because, said she, lice kept coming out of her body. I gave her sulphur baths, and I often tried to ascertain whether there was truth in the assertion, but I always found the skin perfectly clean, and without a rash. Another patient, who was very queer in the head, for six months after the ménopause, had the same notion, and actually put some bugs in a bottle, and used to exhibit them as having come out from different parts of the body. I again bear witness to the admirable effects produced upon cutaneous affections by the external and internal exhibition of the mineral waters of Harrogate and Aix in Savoy.

CHAPTER XII.

OTHER AFFECTIONS OCCURRING AT THE CHANGE OF LIFE IN
500 WOMEN.

TABLE XXXIII.

*Liability of 500 Women to other Affections at the Change
of Life.*

Gout	3
Rheumatic fever	3
Rheumatic joints	7
Hypertrophy of the heart	1
Subacute peritonitis	1
Ascites	1
Consumption, aggravated	3
Bronchitis, aggravated	4
Asthma, aggravated	1
Hæmoptysis	6
Ruptured varicose veins	3
Epistaxis	9
Chronic otorrhœa	1
Total	43

GOUT.—According to Hippocrates, women are not subject
to gout until after cessation; but he alluded to Grecian
women living in retirement. When women lead the life of
men, they become to a certain extent, liable to their disorders;
and therefore Seneca might have been justified in saying of
the women of his time, that they were liable to gout " ob

varii generis debachationes." As far as my experience goes, women are little subject to gout, but I have, like Chomel and Ferrus, observed it at puberty and when menstruation was fully established, and most frequently after the ménopause when it assumed an anomalous form, as noticed by Dr. Gairdner. Even then, it is a disease of rare occurrence, so much so that I do not remember more than four cases amongst women of the upper classes during the last twelve years, subsequent to my collecting from dispensary and other patients the statistics on which this work was originally founded.

The following table, extracted from the Registrar-General's Report, throws light on this subject :—

TABLE XXXIV.

Relative Mortality from Gout in both Sexes at successive Periods.

Ages.	Males.	Females.
5	...	11
5 to 10		
10 ,, 15	...	212
15 ,, 20	67	
20 ,, 30	168	56
30 ,, 40	541	121
40 ,, 50	732	291
50 ,, 60	1148	152
60 ,, 70	458	103
70 ,, 80	186	
Total	3300	946

This shows that women are most liable to gout at puberty and at the change of life ; but before the menstrual flow is regularly established, gout generally assumes an anomalous character, being accompanied by singular nervous symptoms, of which I have seen some remarkable instances. At the change of life, and after cessation, this disease may follow its usual course, and Trousseau has observed that hysterical symptoms may be cut short by an attack of gout. It is curious to note that up to fifteen, no seeds of gout have been evolved in the male sex, whereas many had grown up in the female, and that while in man, the liability to gout goes on

U

increasing from fifteen to sixty, in women, puberty seems to sweep away its tendency, for there is none from fifteen to twenty, though after that age, when the body has attained its full size, the seeds develope themselves, gout becoming most frequent between forty and fifty.

RHEUMATIC FEVER.—The statistics of rheumatic affections likewise evince that from forty to fifty a change takes place in the female constitution, for up to forty more men than women die from rheumatic affections, while from forty to fifty the deaths from these affections are as 604 women to 295 men, and from fifty to sixty as 755 women to 338 men. This increased liability to rheumatism after the ménopause may perhaps depend on the skin being then more given to perspire, and on the greater frequency of checked perspiration. The rheumatism of those who are nursing can be thus accounted for.

ASCITES.—I have seen only two cases of this affection at the change of life, but I have very frequently noticed the pitting, on pressure, of both the lower limbs at this period. On referring to the Registrar-General's Reports, I find that more women, than men, die of dropsy from forty to fifty, in the proportion of nine to five; and from fifty to sixty, in the proportion of fourteen to six. Portal and Gardanne have both stated, that women were particularly subject to dropsical effusions of an obstinate nature at the change of life. Breschet, in his researches on *Hydropsies Actives*, relates the case of a lady, menstruation came at thirteen, and ceased at forty-eight, though the characteristic phenomena of cessation are said to have only appeared at fifty-four; namely, great debility, flushes, vertigo, headache. At fifty-five, after a fit of anger, the whole body swelled considerably, except the arms. There was fever, furred tongue, headache, oppression, and sediments in the urine. Twelve days after this, eighteen leeches were applied to the anus, and the swelling diminished; on the eighteenth day an emetic was given, and on the twenty-eighth all signs of effusion had disappeared. This case brings to my recollection another instance of dropsy, caused by severe and sudden mental perturbation. A gentleman, aged twenty-five, who had habitually enjoyed good health, was standing at the altar and about to be married, when some one stepped forward to forbid the ceremony. On his return home his legs swelled, serum was effused in the cellular tissue, ascites appeared, the urine was albuminous, and in a few months the patient died. A post-mortem examination was not permitted.

CONSUMPTION.—On consulting those practitioners whose experience in pulmonary disorders enables them to speak authoritatively on the subject, I find them of my opinion, that the morbid influence of the change of life on consumption is not great. If consumption appear at this period, it has previously existed under a latent form, as in the three cases noted in my statistics, in which the disease was aggravated by the ménopause, and I have not met with another case during the last twelve years. I must observe, however, that Dubois mentions that two ladies who were saved from phthisis by the regular establishment of the menstrual flow, fell victims to that complaint at the change of life without any other apparent cause. B. de Boismont says, that in ten cases he has been the sad spectator of the rapid progress of consumption at this epoch, which had previously remained stationary for years. More recently Dr. Emmett, of New York, has recorded his conviction that dysmenorrhœa, by causing the early cessation of menstruation, is a frequent cause of phthisis in comparatively young women ; but this assertion does not accord with my own observation.

BRONCHITIS.—In four of my cases, habitual bronchitis dated from the change of life, and seemed to have been caused by it ; and Bordeu has noted similar cases.

HÆMOPTYSIS.—This occurred to eight patients, and was evidently caused by the change of life arresting an habitual hæmorrhage. Bordeu has also observed cessation to be followed by monthly attacks of congestion of the lungs, and by spitting of blood.

HEART-DISEASE.—The heart suffers very little at the change of life; my experience on this point again coincides with that of Dr. Quain, for out of my 500 cases, there was only one of hypertrophy with morbid sounds in a woman, aged sixty, and she presented the only instance of *arcus senilis*.

EPISTAXIS.—This critical effort of nature may become a disease, and the nose may require plugging. Menville gives a case wherein slight epistaxis occurred repeatedly until the dodging time, when it became very abundant until the menstrual flow ceased. I have seen women subject to this complaint for from three to five years after cessation, but never to any great extent.

RUPTURED VARICOSE VEINS.—Stahl has observed this to occur at cessation. I have seen it in three instances, and I have repeatedly been told that after cessation the legs had swelled and were hard, hot, and red, the capillaries of the skin being greatly injected.

HYPERTROPHY OF THE THYROID GLAND.—In three patients, during the last few years, I have seen cessation followed by an increase of the thyroid gland to about treble its usual size. All three patients had long suffered from chronic uterine disease; two got well under the influence of iodine rubbed in and taken internally, and I lost sight of the other.

OTORRHŒA.—I have noticed this complaint to come on periodically with the menstrual flow, and the fact has not escaped Mr. Harvey's notice. M. S., aged fifty-two, suffered much from otorrhœa for two years previous to the first menstrual flow, very little from it while menstruation continued regular, but it returned with renewed pertinacity some months before cessation took place.

SPINAL DEVIATION.—B. de Boismont has related two cases in which very extensive deviation immediately followed the change of life, but in all probability a certain amount of deviation had occurred at puberty, and had remained stationary until the ménopause; and I have known this to happen once.

PTYALISM.—Abundant and long-continued salivation has been noticed by Bouchut as a symptom of cessation, but I have only noticed it to occur at that time in connexion with very intense cerebral neuralgia.

INDEX.

THE END.

OPINIONS OF THE PRESS.

"LE livre du Dr. Tilt a son originalité, et devroit être signalé à ceux qui ne considèrent pas la pathologie des âges comme un hors-d'œuvre à l'usage des médecins littérateurs. Le traité *l'âge critique* est rempli de données intéressantes; l'auteur, qui s'est spécialement occupé des maladies des femmes, et qui a publié sur ce sujet des travaux estimés, y apportait une solide expérience ; il avait le mérite de parler d'après sa pratique et de ne pas seulement rajeunir par une phraséologie nouvelle des citations d'un autre temps. En somme, on aurait peine à trouver sur le même sujet une monographie qui valut celle sur laquelle nous venons d'appeler l'attention. Une preuve entre autres que le livre est d'un vrai mérite c'est qu'il éveille dans le lecteur le désir de le reprendre en sousœuvre, de vérifier ce qui parait juste, de contrôler par une nouvelle étude ce qui semble moins admissible, et surtout de tenir meilleur compte des phénomènes de *l'âge critique* dans l'observation de tous les jours."—*Archives Générales de Médecine*, October, 1858.

"On y trouvera une étude très bien faite et toute nouvelle des réactions nerveuses et réflexes que les désordres ovariques et utérins exercent sur toute la machine sensible de la femme. On comprendra avec lui une foule de manifestations pathologiques qui sont le désespoir du médecin clinique. Cette partie de l'œuvre du médecin anglais suffirait à elle seule pour grandir encore, si cela était possible, la reputation que le docteur Tilt s'est acquis comme practicien des plus distingués, comme savant physiologiste, et pour le placer parmi les écrivains dont les travaux ont fait faire un pas réel aux maladies de la compagne de l'homme. Ajoutons que le livre que nous voudrions faire connaître plus au long, se distingue encore par une profonde érudition, de la richesse dans le style, et une saveur scientifique qu'on ne trouve pas toujours dans les œuvres de ce genre."—*Union Médicale*, August, 1857.

"The volume before us is an expansion of a much smaller one, which made its appearance some years ago. Dr. Tilt has now fairly occupied the field with a work, which, if it does not include all that is known upon the diseases of the climacteric period, is at all events a repertory of information upon a subject not very generally understood. The work is divided into twelve chapters : the first five are, an introductory one on the physiology of the change of life, one on its pathology, one on its therapeutics, and one on its hygienics. Then follow chapters which treat consecutively of the diseases of the reproductive organs at this period of life, of the diseases of the digestive organs, and of the skin ; the tenth treats of the diseases of the ganglionic nervous system, and the eleventh of the cerebro-spinal affections ; and the concluding chapter is miscellaneous. Thus the subject of climacteric derangements is pretty nearly exhausted, and additional value is given to the volume by numerous interesting tables, which exhibit various physiological and pathological facts in a clear and definite manner.

"There is a fund of practical matter in Dr. Tilt's book, and no small share of theory also, which is very clearly enunciated. As the best work on the subject of which it treats, we can cordially recommend it to the profession."—*Lancet*.

UTERINE AND OVARIAN INFLAMMATION,

AND ON THE

Physiology and Diseases of Menstruation.*

Third Edition, with Illustrations, 8vo, cloth, 12s.

"THERE are few works that are more steadily earning the approval of the profession than Dr. Tilt's 'Treatise upon Metritis and Ovaritis.' The author fully deserves that it should be so. He may be said to have made the subject his own, so far as English literature is concerned, and to have delicately, scientifically, and satisfactorily discussed a difficult subject and obscure topic. The third edition is now before us. It is indeed a goodly book; but it could scarcely be less voluminous, seeing how wide a field the author traverses. Very much is actually included in the apparently limited subject discussed by the author, who endeavours to keep the reader *au courant* with the teachings prevailing upon the Continent as well as at home.

"When Dr. Tilt started in practice, twenty-five years since, a firm belief in the infallibility of the speculum and of the nitrate of silver was nearly all that was essential for a specialist in the diseases of the sex who prided himself upon being of the advanced school. Matters have changed since then; for in spite of all said and written upon specula and caustics, women would still persist in being ill, and in vaguely referring their troubles to the pelvis. The truth is, as the author remarks, that above the internal sphincter, beyond which neither caustics nor specula can reach, there remains the body of the womb, with its lining membrane, and the ovaries, which for thirty years are thrown into a state of hæmorrhagic and other orgasm every month for several days. Obstetricians have of late years recognised this, and not refused to touch where they could not see, and to use remedies, though they could not apply caustics. For much of the advance that has been made within the last ten years in this department, we are indebted to the exertions of Dr. Tilt. The improvement which the present edition has undergone will tend to propagate the author's doctrines. That they will be in the main fully established we have no doubt. This their truthfulness will do for them; but, independently of this, the clear and simple style of writing in which they have been conveyed must materially contribute to their wide dissemination. What is easy and agreeable to read will be read accordingly. If a man write crabbedly, his readers become crabbed too, and are apt to let him rest quietly on their shelves. Dr. Tilt is a model author, too, in this: that he supplies us with a capital table of contents, a good general index, and an index of bibliography. These may appear trifles to some authors who utterly neglect them. But to all readers they are of great importance, as affording facility for reference. To notice again in detail a work which we have twice before submitted to critical review is quite unnecessary. The treatise

---

* London: John Churchill and Sons, New Burlington Street.

before us now takes its position by the side of our standard text-books, and is too well appreciated to need any further recommendation from ourselves.

" To those who may be called upon to treat diseases of the female generative organs, we know of no work which they may study with greater advantage than this of Dr. Tilt; the author has shown himself to be an original, painstaking, and accurate observer, and no one, however much they may be inclined to differ with him on certain points herein discussed, can say that this work is not a credit to his skill and sagacity in unravelling many very intricate and difficult questions in ovarian and uterine pathology. In corroboration of our opinion we need only point to the fact that this is the third edition which has appeared in comparatively a few years.

" We regret very much that our space does not allow of a more lengthy examination of this most interesting portion of the work, and especially the last chapter relating to the pathology and treatment of hæmatocele, a disease to which the author was the first in this country to direct the attention of the profession, and which has lately, from the researches of Bernutz and Goupil, attracted considerable notice. We have said enough, however, to interest our readers in this work, and we can promise them a vast accession of knowledge of a most useful kind if they will attentively study it."—*London Medical Review.*

" We have on former occasions felt it our duty to remark upon the labours of Dr. Tilt with satisfaction, and as this is the third edition published, we have but little more to say than that he fully maintains the credit before accorded him for industry, and has added to the knowledge we possessed on several important branches of inquiry."—*Medico-Chirurgical Review.*

" Evidently the work of a man of industry, who, as student and practitioner, has devoted immense pains to the investigation of the whole subject of the laws of woman's life in health and disease."—*Medical Times and Gazette.*

" We look upon Dr. Tilt's as one of the really genuine works of the present day."—*Dublin Medical Press.*

" Well deserves perusal."—*Edinburgh Medical Journal.*

" We have long had on our table the fourth edition of Bennet's book 'On Inflammation of the Womb,' and the third of Tilt's 'On Uterine and Ovarian Inflammation,' works which had served to revolutionize uterine pathology and therapeutics, not, it is true, without eliciting much opposition on account of their adopting a too exclusive pathological theory, but which works must ever mark an era in our knowledge of uterine disease."—*Dublin Quarterly Journal of Medical Science.*

" Le Dr. Tilt a introduit dans la science une idée qui depuis lors a porté fruits, et qui, lorsqu'il l'énonçait dogmatiquement en 1848, semblait plutôt une assertion ingénieuse qu'une vérité probable. La pathologie utérine, concentrée presqu' exclusivement dans les lésions du col de l'utérus, laissait inexpliqués un certain nombre de phénomènes d'une observation fréquente. Le Dr. Tilt montra qu'on n'avait envisagé qu'un côté de la pathologie en n'examinant qu'une portion de l'organe, et il prit à tâche d'assigner une juste place aux affections du corps de l'utérus, du péritoine, et surtout des ovaires. Par une réaction facile à pressentir, le col de l'uterus fut peut-être relégué trop loin sur les derniers plans, mais on y gagna d'accorder enfin aux maladies de l'ovaire une attention qu'on leur avait refusé jusque-là."—*Archives Générales de Médecine.*

HANDBOOK OF UTERINE THERAPEUTICS.

"For many years past Dr. Tilt has sought to determine the real value of those various modes of treating inflammatory affections of the womb which have been advocated by different eminent practitioners. In giving the result of his labours to the profession, the author has done a good work. The present volume contains a great deal of useful matter; and though there are some observations in it which do not accord with our own views, yet we strongly recommend our readers to peruse Dr. Tilt's volume for themselves instead of trusting to any critical remarks. They will find its pages very interesting, and at the end of their task will feel grateful to the author for many very valuable suggestions as to the treatment of uterine disease. The different views advocated are illustrated by the notes of many curious cases, so that the reader will find no difficulty in fixing his attention to the end of the work."—*Lancet.*

"But it is fair to say that there is a large amount of useful information scattered throughout the book, available for those cases in which local treatment is useful, and that we thoroughly believe in Dr. Tilt's good faith. The passages we have quoted are proofs of the writer's sincerity, and all details respecting injections, suppositories, incisions, and dilatations, pessaries, leeches, diet, &c., are here to be found in abundance."—*Medical Times.*

"Dr. Tilt has devoted many years to the practical study of the uterus, and anything that comes from his pen upon this subject deserves attention. In the present work he passes in review the various therapeutical measures, both medical and surgical, which are adapted to relieve the diseases of the organs peculiar to females. His style is lucid, and his observations are those of an intelligent observer and thoughtful practitioner. Considering the very large number of cases of uterine disturbance which come under the notice of the profession, a work which treats specially of such matters must possess very general interest, and Dr. Tilt's work is very creditable to his research and to his practical skill, and it will be an acceptable handbook on the shelves of medical practitioners."—*Medical Circular.*

"Dr. Tilt's 'Handbook of Uterine Therapeutics,' supplies a want which has often been felt, by bringing together, within moderate compass, the therapeutic agents which are most serviceable in the treatment of the diseases of women ; so that the practitioner, with this little work in his hand, is enabled in a short space of time to determine what remedies will best suit the requirements of any individual case. It may, therefore, not only be read with pleasure and instruction, but will also be found very useful as a book of reference. A formulary is appended, in which are contained some useful prescriptions, both for internal and external use. Glycerine enters largely into the latter class of compounds, and the preparations in which it serves the purpose of a vehicle will be found preferable to the ointments generally employed, as its demulcent, detergent, and absorbefacient properties render it very superior to lard and the other excipients which are commonly used."—*The Medical Mirror.*

" This volume will be found a safe guide to the practitioner, and is concluded by a carefully selected formulary of the preparations which in the author's experience have been most beneficial. Each subject is handled with the delicacy required by the nature of the question alluded to, and is illustrated by a brief but sufficient account of cases from the author's practice, and we much regret our inability to bestow more than a passing glance at so valuable a contribution to uterine pathology."—*Journal of Practical Medicine and Surgery.*

" We cannot but admire the evident sincerity of the author of this little work. It is simply what it pretends to be, 'A Handbook of Uterine Therapeutics,' and we make no doubt that many members of our profession will thank Dr. Tilt for furnishing them with so well-arranged a collection of therapeutical resources. There is certainly a large amount of useful information to be found in the book, and we recommend it to those who wish to have ready at hand, in a condensed form, a fair account of all the *armamentaria* which may be arranged and brought into play for the relief or cure of some of the most painful affections which distress and render unhappy those who demand our warmest sympathy and our highest respect as

' The most faultless of created things.' "

—*Madras Quarterly Journal of Medical Science.*

ELEMENTS OF HEALTH,

AND

PRINCIPLES OF HYGIENE.*

"THERE are two kinds of popular medical writers—those who introduce the public into the sanctuary of medical science, and tempt them to poison themselves by injudiciously taking medicines; and those who seek to improve the sanitary state of mankind by diffusing a knowledge of the general laws which govern nature, in relation to living creatures, and by imparting those precepts of physiology which, if duly observed, would prevent disease. The first class of writers we heartily condemn. To illustrate the second we point to the names of Drs. James Johnson, Mayo, and particularly Dr. A. Combe, deeming them benefactors of the human race. Following in the footsteps of those just mentioned, is Dr. Tilt. In his "Elements of Health" he has successfully done for women what the others have done for men, and his work is a model for those who propose writing on similar subjects, for in a vast plan every subject receives comment in proportion to its importance, and is lucidly explained so as to bring conviction to every woman of ordinary capacity. The work is characterized by extreme delicacy of expression, a healthy tone of feeling, free from all mawkish leaning to the prejudices of the sex, and it is written in a style which rivets the attention and carries on the reader from page to page. Our space is claimed by professional subjects, so that we cannot review this book so completely as we could wish. We can merely trace its general plan and prevailing idea. Each successive period of seven years forms a chapter, in which the mental and moral progress of decay are sketched, while the physical is treated at full length. Food, sleep, exercise, clothing, occupations, are separately considered; and the chapter concludes with a brief account of the diseases which are common to each epoch, and of the indications heralding their approach, which render medical advice imperative. Dr. Tilt's prevailing idea seems to be, that further improvement in the sanitary condition of society is to be principally effected by giving women an insight into the laws to which they are subjected, as living beings and as women; their own health, the improvement of the human race, and the welfare of society being attainable by that means. The work seems, also, to commend itself to the profession by the careful manner in which is therein laid down the means of preventing that exaggeration of the nervous temperament which is so fruitful a source of the diseases of women. In conclusion, we shall only add, that as Dr. Tilt's is the only work of the kind—at least in English literature—we trust it will be considered an indispensable guide by persons to whom may be entrusted the sacred task of educating the present generation of children, who are necessarily to become our future generations of men and women."—*Lancet.*

---

* London: Bohn, York Street, Covent Garden.

"In the *British and Foreign Medico-Chirurgical Quarterly Review* it was lately remarked that a treatise on female hygiene was much wanted; and all those engaged in general practice who have to contend daily with the ignorance and prejudices of women respecting themselves and their children, will re-echo the assertion of our respected contemporary. Dr. Tilt has sought to fill up this desideratum; and we are anxious to be among the first to notice a book which originated in our columns. Two years ago Dr. Tilt inserted in this journal some highly interesting papers on the right management of women at the critical periods of life. These papers have suggested to the author the present work, of which we intend briefly to sketch the outline. The work is divided into periods of seven years, and each period forms a chapter. Each chapter briefly notices the mental and moral development or decay, and the physical condition is treated with care. The food, clothing, exercise, and sleep, as regards each epoch, are passed in review; and the diseases to which women at each period are most liable are pointed out, as well as the most appropriate means of prevention. Every chapter is preceded and followed by tables showing the mortality of both sexes for each year successively, the mean duration of life, and its value for insurance purposes—calculations which derive importance from the fact of their having been made under the eye of Mr. Farr, of the Registrar-General's office. Such is the outline of a work which combines a vast amount of information in a small compass, and of which we regret that our space will not allow us to give extracts; it is much required, and will doubtless ere long become as popular as those of the late lamented Dr. Combe. Perhaps no man is better calculated than Dr. Tilt to fill up this hiatus in medical literature; for few unite to the same extent great opportunities of observation with sterling common sense, a thorough love of his subject, and a lucid, correct, and lively style. We think the work will be found as useful to the practitioner as it is indispensable to those who are in any way connected with the education or responsibilities of women; for while, on the one hand, it is the best treatise on physical education with which we are acquainted, it also affords practitioners excellent advice respecting the prevention of nervous complaints, and, in fact, of all the diseases to which women are amenable from the peculiarities of their formation and habits."—*Provincial Medical and Surgical Journal.*

"Dr. Tilt has chosen a subject which required great tact and delicacy for its treatment; and though such a work was much wanted, it has been this feeling probably which has deterred writers from entering on the field before. We think Dr. Tilt has succeeded. He has taken up most carefully all those departments of statistical inquiry which throw light on the differences that exist in the constitution and temperament of the sexes, and in all parts of his work has treated the subject in both a learned and a practical manner."—*Athenæum.*

London, New Burlington Street,
October, 1870.

MESSRS. CHURCHILL & SONS'

Publications,

IN

MEDICINE

AND THE VARIOUS BRANCHES OF

NATURAL SCIENCE.

"It would be unjust to conclude this notice without saying a few words in favour of Mr. Churchill, from whom the profession is receiving, it may be truly said, the most beautiful series of Illustrated Medical Works which has ever been published."—*Lancet.*

"All the publications of Mr. Churchill are prepared with so much taste and neatness, that it is superfluous to speak of them in terms of commendation."—*Edinburgh Medical and Surgical Journal.*

"No one is more distinguished for the elegance and *recherché* style of his publications than Mr. Churchill."—*Provincial Medical Journal.*

"The name of Churchill has long been a guarantee for the excellence of illustrated works, and it would be superfluous to repeat the admiration that we have several times expressed in this respect, of the spirit with which this firm engages in these costly but valuable series."—*Medical Press and Circular.*

"The typography, illustrations, and getting up are, in all Mr. Churchill's publications, most beautiful."—*Monthly Journal of Medical Science.*

"Mr. Churchill's illustrated works are among the best that emanate from the Medical Press."—*Medical Times.*

"We have before called the attention of both students and practitioners to the great advantage which Mr. Churchill has conferred on the profession, in the issue, at such a moderate cost, of works so highly creditable in point of artistic execution and scientific merit."—*Dublin Quarterly Journal.*

Messrs. Churchill & Sons are the Publishers of the following Periodicals, offering to Authors a wide extent of Literary Announcement, and a Medium of Advertisement, addressed to all Classes of the Profession.

THE BRITISH AND FOREIGN MEDICO-CHIRURGICAL REVIEW, AND QUARTERLY JOURNAL OF PRACTICAL MEDICINE AND SURGERY.
Price Six Shillings. No. XCII.

THE QUARTERLY JOURNAL OF MICROSCOPICAL SCIENCE,
Edited by DR. LANKESTER, F.R.S., and E. RAY LANKESTER, B.A., F.R.M.S. Price 4s.
No. XL. *New Series.*

THE JOURNAL OF MENTAL SCIENCE.
By authority of the Medico-Psychological Association.
Edited by HENRY MAUDSLEY, M.D.
Published Quarterly, price 3s. 6d. *New Series.*
No. XXXIX.

JOURNAL OF CUTANEOUS MEDICINE.
Edited by Dr. H. S. PURDON, Belfast.
Published Quarterly, price 2s. No. XIV.

ARCHIVES OF MEDICINE:
A Record of Practical Observations and Anatomical and Chemical Researches, connected with the Investigation and Treatment of Disease. Edited by Dr. LIONEL S. BEALE, F.R.S. Published Quarterly; Nos. I. to VIII., 3s. 6d.; IX. to XII., 2s. 6d., XIII. to XVI., 3s.

THE ROYAL LONDON OPHTHALMIC HOSPITAL REPORTS, AND JOURNAL OF OPHTHALMIC MEDICINE AND SURGERY.
Vol. VI., Part 4, 2s. 6d.

THE MEDICAL TIMES & GAZETTE.
Published Weekly, price Sixpence.
Annual Subscription, £1. 6s., and regularly forwarded to all parts of the Kingdom, post free, for £1. 8s.

THE PHARMACEUTICAL JOURNAL,
AND TRANSACTIONS OF THE PHARMACEUTICAL SOCIETY.
Published Weekly, price Fourpence.
₄ May also be had in Monthly Parts.

THE BRITISH JOURNAL OF DENTAL SCIENCE.
Published Monthly, price One Shilling.
No. CLXXII.

THE MEDICAL DIRECTORY.
Published Annually. 8vo. cloth, 10s. 6d.

THE HALF-YEARLY ABSTRACT OF THE MEDICAL SCIENCES.

BEING A DIGEST OF BRITISH AND CONTINENTAL MEDICINE,

AND OF THE PROGRESS OF MEDICINE AND THE COLLATERAL SCIENCES.

Edited by W. DOMETT STONE, M.D., F.R.C.S., L.S.A.

Post 8vo. cloth, 6s. 6d. Vols. I. to LI.

"American physicians may be congratulated that they are once more favoured with the reprint of 'Ranking's Abstract.' If any doctor is so busy that he can read but a single volume a year, then, assuredly, he should make this his book; for here are collected and condensed the most valuable contributions to periodical medical literature—French, German, British, and American—for the year; and, on the other hand, no physician—it matters not how wide the range of his reading—can fail to find, in this volume, truths that will enlarge his medical knowledge, and precepts that will help him in some of his daily professional needs."—*Cincinnati Journal of Medicine,* April, 1867.

"We have only space to say that this volume is rich in valuable articles, among which there are many on materia medica and therapeutics. Gathered from all sources in the new books and medical journals of Europe and America, this work may be viewed as the cream of that class of medical essays, and is a useful occupant of the physician's office-table, to keep him reminded of the progress of medicine."—*American Journal of Pharmacy,* May, 1867.

A CLASSIFIED INDEX

TO

MESSRS. CHURCHILL & SONS' CATALOGUE.

TO BE COMPLETED IN TWELVE PARTS, 4TO., AT 7s. 6d. PER PART.

PARTS I. & II. NOW READY.

A DESCRIPTIVE TREATISE

ON THE

NERVOUS SYSTEM OF MAN,

WITH THE MANNER OF DISSECTING IT.

By LUDOVIC HIRSCHFELD,

DOCTOR OF MEDICINE OF THE UNIVERSITIES OF PARIS AND WARSAW, PROFESSOR OF ANATOMY TO THE FACULTY OF MEDICINE OF WARSAW;

Edited in English (from the French Edition of 1866)

By ALEXANDER MASON MACDOUGAL, F.R.C.S.,

WITH

AN ATLAS OF ARTISTICALLY-COLOURED ILLUSTRATIONS,

Embracing the Anatomy of the entire Cerebro-Spinal and Sympathetic Nervous Centres and Distributions in their accurate relations with all the important Constituent Parts of the Human Economy, and embodied in a series of 56 Single and 9 Double Plates, comprising 197 Illustrations,

Designed from Dissections prepared by the Author, and Drawn on Stone by

J. B. LÉVEILLÉ.

WILLIAM ACTON, M.R.C.S.

I.

A PRACTICAL TREATISE ON DISEASES OF THE URINARY
AND GENERATIVE ORGANS IN BOTH SEXES. Third Edition. 8vo. cloth,
£1. 1s. With Plates, £1. 11s. 6d. The Plates alone, limp cloth, 10s. 6d.

II.

THE FUNCTIONS AND DISORDERS OF THE REPRODUC-
TIVE ORGANS IN CHILDHOOD, YOUTH, ADULT AGE, AND ADVANCED
LIFE, considered in their Physiological, Social, and Moral Relations. Fourth Edition.
8vo. cloth, 10s. 6d.

III.

PROSTITUTION : Considered in its Moral, Social, and Sanitary Aspects,
Second Edition, enlarged. 8vo. cloth, 12s.

ROBERT ADAMS, A.M., C.M., M.D.

A TREATISE ON RHEUMATIC GOUT ; OR, CHRONIC
RHEUMATIC ARTHRITIS. 8vo. cloth, with a Quarto Atlas of Plates, 21s.

WILLIAM ADAMS, F.R.C.S.

I.

ON THE PATHOLOGY AND TREATMENT OF LATERAL
AND OTHER FORMS OF CURVATURE OF THE SPINE. With Plates.
8vo. cloth, 10s. 6d.

II.

CLUBFOOT : its Causes, Pathology, and Treatment. Jacksonian Prize Essay
for 1864. With 100 Engravings. 8vo. cloth, 12s.

III.

ON THE REPARATIVE PROCESS IN HUMAN TENDONS
AFTER SUBCUTANEOUS DIVISION FOR THE CURE OF DEFORMITIES.
With Plates. 8vo. cloth, 6s.

IV.

SKETCH OF THE PRINCIPLES AND PRACTICE OF
SUBCUTANEOUS SURGERY. 8vo. cloth, 2s. 6d.

WILLIAM ADDISON, F.R.C.P., F.R.S.

I.

CELL THERAPEUTICS. 8vo. cloth, 4s.

II.

ON HEALTHY AND DISEASED STRUCTURE, AND THE TRUE
PRINCIPLES OF TREATMENT FOR THE CURE OF DISEASE, ESPECIALLY CONSUMPTION
AND SCROFULA, founded on MICROSCOPICAL ANALYSIS. 8vo. cloth, 12s.

C. J. B. ALDIS, M.D., F.R.C.P.

AN INTRODUCTION TO HOSPITAL PRACTICE IN VARIOUS
COMPLAINTS ; with Remarks on their Pathology and Treatment. 8vo. cloth, 5s. 6d.

SOMERVILLE SCOTT ALISON, M.D.EDIN., F.R.C.P.

THE PHYSICAL EXAMINATION OF THE CHEST IN PUL-
MONARY CONSUMPTION, AND ITS INTERCURRENT DISEASES. With
Engravings. 8vo. cloth, 12s.

JULIUS ALTHAUS, M.D., M.R.C.P.

ON EPILEPSY, HYSTERIA, AND ATAXY. Cr. 8vo. cloth, 4s.

THE ANATOMICAL REMEMBRANCER; OR, COMPLETE

POCKET ANATOMIST. Sixth Edition, carefully Revised. 32mo. cloth, 3s. 6d.

McCALL ANDERSON, M.D., F.F.P.S.

I.

THE PARASITIC AFFECTIONS OF THE SKIN. Second

Edition. With Engravings. 8vo. cloth, 7s. 6d.

II.

ECZEMA. Second Edition. 8vo. cloth, 6s.

III.

PSORIASIS AND LEPRA. With Chromo-lithograph. 8vo. cloth, 5s.

J. T. ARLIDGE, M.D.LOND., F.R.C.P.

ON THE STATE OF LUNACY AND THE LEGAL PROVISION

FOR THE INSANE; with Observations on the Construction and Organisation of Asylums. 8vo. cloth, 7s.

GEORGE ARMATAGE, M.R.C.V.S.

THE VETERINARIAN'S POCKET REMEMBRANCER: con-

taining concise directions for the Treatment of Urgent or Rare Cases, embracing Semeiology, Diagnosis, Prognosis, Surgery, Therapeutics, Detection of Poisons, Hygiene, &c. Post 18mo., 3s.

ALEXANDER ARMSTRONG, M.D., F.R.C.P., R.N.

OBSERVATIONS ON NAVAL HYGIENE AND SCURVY.

More particularly as the latter appeared during a Polar Voyage. 8vo. cloth, 5s.

T. J. ASHTON, M.R.C.S.

I.

ON THE DISEASES, INJURIES, AND MALFORMATIONS

OF THE RECTUM AND ANUS. Fourth Edition. 8vo. cloth, 8s.

II.

PROLAPSUS, FISTULA IN ANO, AND OTHER DISEASES

OF THE RECTUM; their Pathology and Treatment. Third Edition. Post 8vo. cloth, 3s. 6d.

THOS. J. AUSTIN, M.R.C.S.ENG.

A PRACTICAL ACCOUNT OF GENERAL PARALYSIS:

Its Mental and Physical Symptoms, Statistics, Causes, Seat, and Treatment. 8vo. cloth, 6s.

A. W. BARCLAY, M.D., F.R.C.P.

I.

A MANUAL OF MEDICAL DIAGNOSIS. Second Edition.

Foolscap 8vo. cloth, 8s. 6d.

II.

MEDICAL ERRORS.—Fallacies connected with the Application of the

Inductive Method of Reasoning to the Science of Medicine. Post 8vo. cloth, 5s.

III.

GOUT AND RHEUMATISM IN RELATION TO DISEASE

OF THE HEART. Post 8vo. cloth, 5s.

G. H. BARLOW, M.D., F.R.C.P.

A MANUAL OF THE PRACTICE OF MEDICINE. Second Edition. Fcap. 8vo. cloth, 12s. 6d.

ROBERT BARNES, M.D., F.R.C.P.

LECTURES ON OBSTETRIC OPERATIONS, INCLUDING THE TREATMENT OF HÆMORRHAGE, and forming a Guide to the Management of Difficult Labour. With nearly 100 Engravings. 8vo. cloth, 15s.

E. BASCOME, M.D.

A HISTORY OF EPIDEMIC PESTILENCES, FROM THE EARLIEST AGES. 8vo. cloth, 8s.

W. R. BASHAM, M.D., F.R.C.P.

I.

RENAL DISEASES; a CLINICAL GUIDE to their DIAGNOSIS and TREATMENT. 8vo. cloth, 7s.

II.

ON DROPSY, AND ITS CONNECTION WITH DISEASES OF THE KIDNEYS, HEART, LUNGS AND LIVER. With 16 Plates. Third Edition. 8vo. cloth, 12s. 6d.

FREDERIC BATEMAN, M.D., M.R.C.P.

APHASIA OR LOSS OF SPEECH, and the LOCALISATION of the FACULTY of ARTICULATE LANGUAGE. 8vo., 7s.

H. F. BAXTER, M.R.C.S.L.

ON ORGANIC POLARITY; showing a Connexion to exist between Organic Forces and Ordinary Polar Forces. Crown 8vo. cloth, 5s.

LIONEL J. BEALE, M.R.C.S.

I.

HEALTH AND LONGEVITY. Second Edition. Foolscap 8vo., 3s. 6d.

II.

THE LAWS OF HEALTH IN THEIR RELATIONS TO MIND AND BODY. A Series of Letters from an Old Practitioner to a Patient. Post 8vo. cloth, 7s. 6d.

LIONEL S. BEALE, M.B., F.R.S., F.R.C.P.

I.

ON KIDNEY DISEASES, URINARY DEPOSITS, AND CALCULOUS DISORDERS. Third Edition, much Enlarged. With 70 Plates. 8vo. cloth, 25s.

II.

THE MICROSCOPE, IN ITS APPLICATION TO PRACTICAL MEDICINE. Third Edition. With 58 Plates. 8vo. cloth, 16s.

III.

PROTOPLASM; OR, LIFE, MATTER AND MIND. Second Edition. With 8 Plates. Crown 8vo. cloth, 6s. 6d.

IV.

DISEASE GERMS; their SUPPOSED NATURE. An ORIGINAL INVESTIGATION. With Plates. Crown 8vo., 3s. 6d.

HENRY BEASLEY.

I.

THE BOOK OF PRESCRIPTIONS; containing 3000 Prescriptions.

Collected from the Practice of the most eminent Physicians and Surgeons, English and Foreign. Third Edition. 18mo. cloth, 6s.

II.

THE DRUGGIST'S GENERAL RECEIPT-BOOK: comprising a

copious Veterinary Formulary and Table of Veterinary Materia Medica; Patent and Proprietary Medicines, Druggists' Nostrums, &c.; Perfumery, Skin Cosmetics, Hair Cosmetics, and Teeth Cosmetics; Beverages, Dietetic Articles, and Condiments; Trade Chemicals, Miscellaneous Preparations and Compounds used in the Arts, &c.; with useful Memoranda and Tables. Sixth Edition. 18mo. cloth, 6s.

III.

THE POCKET FORMULARY AND SYNOPSIS OF THE

BRITISH AND FOREIGN PHARMACOPŒIAS; comprising standard and approved Formulæ for the Preparations and Compounds employed in Medical Practice. Eighth Edition, corrected and enlarged. 18mo. cloth, 6s.

HENRY BENNET, M.D.

I.

A PRACTICAL TREATISE ON UTERINE DISEASES.

Fourth Edition, revised, with Additions. 8vo. cloth, 16s.

II.

WINTER AND SPRING ON THE SHORES OF THE MEDI-

TERRANEAN: OR, THE RIVIERA, MENTONE, ITALY, CORSICA, SICILY, ALGERIA, SPAIN, AND BIARRITZ, AS WINTER CLIMATES. Fourth Edition, with numerous Plates, Maps, and Wood Engravings. Post 8vo. cloth, 12s.

ROBERT BENTLEY, F.L.S.

A MANUAL OF BOTANY. With nearly 1,200 Engravings on Wood.

Second Edition. Fcap. 8vo. cloth, 12s. 6d.

ALBERT J. BERNAYS, PH.D., F.C.S.

NOTES FOR STUDENTS IN CHEMISTRY; being a Syllabus com-

piled from the Manuals of Miller, Fownes, Berzelius, Gerhardt, Gorup-Besanez, &c. Fifth Edition. Fcap. 8vo. cloth, 3s. 6d.

HENRY HEATHER BIGG.

ORTHOPRAXY: a complete Guide to the Modern Treatment of Deformi-

ties by Mechanical Appliances. With 300 Engravings. Second Edition. Post 8vo. cloth, 10s.

S. B. BIRCH M.D., M.R.C.P.

I.

OXYGEN: ITS ACTION, USE, AND VALUE IN THE TREATMENT

OF VARIOUS DISEASES OTHERWISE INCURABLE OR VERY INTRACTABLE. Second Edition. Post 8vo. cloth, 3s. 6d.

II.

CONSTIPATED BOWELS: the Various Causes and the Different Means

of Cure. Third Edition. Post 8vo. cloth, 3s. 6d.

GOLDING BIRD, M.D., F.R.S.

URINARY DEPOSITS; THEIR DIAGNOSIS, PATHOLOGY,
AND THERAPEUTICAL INDICATIONS. With Engravings. Fifth Edition.
Edited by E. LLOYD BIRKETT, M.D. Post 8vo. cloth, 10s. 6d.

JOHN BISHOP, F.R.C.S., F.R.S.
I.

ON DEFORMITIES OF THE HUMAN BODY, their Pathology
and Treatment. With Engravings on Wood. 8vo. cloth, 10s.

II.

ON ARTICULATE SOUNDS, AND ON THE CAUSES AND
CURE OF IMPEDIMENTS OF SPEECH. 8vo. cloth, 4s.

BLAINE.

OUTLINES OF THE VETERINARY ART; OR, A TREATISE
ON THE ANATOMY, PHYSIOLOGY, AND DISEASES OF THE HORSE,
NEAT CATTLE, AND SHEEP. Seventh Edition. By Charles Steel, M.R.C.V.S.L.
With Plates. 8vo. cloth, 18s.

C. L. BLOXAM.
I.

CHEMISTRY, INORGANIC AND ORGANIC; with Experiments
and a Comparison of Equivalent and Molecular Formulæ. With 276 Engravings on Wood.
8vo. cloth, 16s.

II.

LABORATORY TEACHING; OR PROGRESSIVE EXERCISES
IN PRACTICAL CHEMISTRY. With 89 Engravings. Crown, 8vo. cloth, 5s. 6d.

HONORÉ BOURGUIGNON, M.D.

ON THE CATTLE PLAGUE; OR, CONTAGIOUS TYPHUS IN
HORNED CATTLE: its History, Origin, Description, and Treatment. Post 8vo. 5s.

JOHN E. BOWMAN, & C. L. BLOXAM.
I.

PRACTICAL CHEMISTRY, including Analysis. With numerous Illustrations on Wood. Fifth Edition. Foolscap 8vo. cloth, 6s. 6d.

II.

MEDICAL CHEMISTRY; with Illustrations on Wood. Fourth Edition,
carefully revised. Fcap. 8vo. cloth, 6s. 6d.

P. MURRAY BRAIDWOOD, M.D. EDIN.

ON PYÆMIA, OR SUPPURATIVE FEVER: the Astley Cooper
Prize Essay for 1868. With 12 Plates. 8vo. cloth, 10s. 6d.

JAMES BRIGHT, M.D.

ON DISEASES OF THE HEART, LUNGS, & AIR PASSAGES;
with a Review of the several Climates recommended in these Affections. Third Edition. Post 8vo. cloth, 9s.

WILLIAM BRINTON, M.D., F.R.S.

I.

THE DISEASES OF THE STOMACH, with an Introduction on its Anatomy and Physiology; being Lectures delivered at St. Thomas's Hospital. Second Edition. 8vo. cloth, 10s. 6d.

II.

INTESTINAL OBSTRUCTION. Edited by Dr. BUZZARD. Post 8vo. cloth, 5s.

BERNARD E. BRODHURST, F.R.C.S.

I.

CURVATURES OF THE SPINE: their Causes, Symptoms, Pathology, and Treatment. Second Edition. Roy. 8vo. cloth, with Engravings, 7s. 6d.

II.

ON THE NATURE AND TREATMENT OF CLUBFOOT AND ANALOGOUS DISTORTIONS involving the TIBIO-TARSAL ARTICULATION. With Engravings on Wood. 8vo. cloth, 4s. 6d.

III.

PRACTICAL OBSERVATIONS ON THE DISEASES OF THE JOINTS INVOLVING ANCHYLOSIS, and on the TREATMENT for the RESTORATION of MOTION. Third Edition, much enlarged, 8vo. cloth, 4s. 6d.

CHARLES BROOKE, M.A., M.B., F.R.S.

ELEMENTS OF NATURAL PHILOSOPHY. Based on the Work of the late Dr. Golding Bird. Sixth Edition. With 700 Engravings. Fcap. 8vo. cloth, 12s. 6d.

T. L. BRUNTON, B.SC., M.B.

ON DIGITALIS. With some Observations on the Urine. Fcap. 8vo. cloth, 4s. 6d.

THOMAS BRYANT, F.R.C.S.

I.

ON THE DISEASES AND INJURIES OF THE JOINTS. CLINICAL AND PATHOLOGICAL OBSERVATIONS. Post 8vo. cloth, 7s. 6d.

II.

CLINICAL SURGERY. Parts I. to VII. 8vo., 3s. 6d. each.

FLEETWOOD BUCKLE, M.D., L.R.C.P.LOND.

VITAL AND ECONOMICAL STATISTICS OF THE HOSPITALS, INFIRMARIES, &c., OF ENGLAND AND WALES. Royal 8vo. 5s.

JOHN CHARLES BUCKNILL, M.D., F.R.C.P., F.R.S., & DANIEL H. TUKE, M.D.

A MANUAL OF PSYCHOLOGICAL MEDICINE: containing the History, Nosology, Description, Statistics, Diagnosis, Pathology, and Treatment of Insanity. Second Edition. 8vo. cloth, 15s.

GEORGE BUDD, M.D., F.R.C.P., F.R.S.

I.

ON DISEASES OF THE LIVER. Illustrated with Coloured Plates and Engravings on Wood. Third Edition. 8vo. cloth, 16s.

II.

ON THE ORGANIC DISEASES AND FUNCTIONAL DIS- ORDERS OF THE STOMACH. 8vo. cloth, 9s.

G. W. CALLENDER, F.R.C.S.

FEMORAL RUPTURE: Anatomy of the Parts concerned. With Plates.
8vo. cloth, 4s.

JOHN M. CAMPLIN, M.D., F.L.S.

ON DIABETES, AND ITS SUCCESSFUL TREATMENT.
Third Edition, by Dr. Glover. Fcap. 8vo. cloth, 3s. 6d.

ROBERT B. CARTER, F.R.C.S.

ON THE INFLUENCE OF EDUCATION AND TRAINING
IN PREVENTING DISEASES OF THE NERVOUS SYSTEM. Fcap. 8vo., 6s.

W. B. CARPENTER, M.D., F.R.S.

I.
PRINCIPLES OF HUMAN PHYSIOLOGY. With nearly 300 Illustrations on Steel and Wood. Seventh Edition. Edited by Mr. HENRY POWER. 8vo. cloth, 28s.

II.
A MANUAL OF PHYSIOLOGY. With 252 Illustrations on Steel and Wood. Fourth Edition. Fcap. 8vo. cloth, 12s. 6d.

III.
THE MICROSCOPE AND ITS REVELATIONS. With more than 400 Engravings on Steel and Wood. Fourth Edition. Fcap. 8vo. cloth, 12s. 6d.

JOSEPH PEEL CATLOW, M.R.C.S.

ON THE PRINCIPLES OF ÆSTHETIC MEDICINE; or the Natural Use of Sensation and Desire in the Maintenance of Health and the Treatment of Disease. 8vo. cloth, 9s.

T. K. CHAMBERS, M.D., F.R.C.P.

I.
LECTURES, CHIEFLY CLINICAL. Fourth Edition. 8vo. cloth, 14s.

II.
THE INDIGESTIONS OR DISEASES OF THE DIGESTIVE
ORGANS FUNCTIONALLY TREATED. Second Edition. 8vo. cloth, 10s. 6d.

III.
SOME OF THE EFFECTS OF THE CLIMATE OF ITALY.
Crown 8vo. cloth, 4s. 6d.

H. T. CHAPMAN, F.R.C.S.

I.
THE TREATMENT OF OBSTINATE ULCERS AND CUTA-
NEOUS ERUPTIONS OF THE LEG WITHOUT CONFINEMENT. Third Edition. Post 8vo. cloth, 3s. 6d.

II.
VARICOSE VEINS: their Nature, Consequences, and Treatment, Palliative and Curative. Second Edition. Post 8vo. cloth, 3s. 6d.

JOHN CHAPMAN, M.D., M.R.C.P.

THE MEDICAL INSTITUTIONS OF THE UNITED KING-
DOM; a History exemplifying the Evils of Over-Legislation. 8vo. cloth, 3s. 6d.

PYE HENRY CHAVASSE, F.R.C.S.

I.

ADVICE TO A MOTHER ON THE MANAGEMENT OF HER CHILDREN. Tenth Edition. Foolscap 8vo., 2s. 6d.

II.

COUNSEL TO A MOTHER: being a Continuation and the Completion of "Advice to a Mother." Fcap. 8vo. 2s. 6d.

III.

ADVICE TO A WIFE ON THE MANAGEMENT OF HER OWN HEALTH. With an Introductory Chapter, especially addressed to a Young Wife. Ninth Edition. Fcap. 8vo., 2s. 6d.

F. LE GROS CLARK, F.R.C.S.

I.

LECTURES ON THE PRINCIPLES OF SURGICAL DIAGNOSIS : ESPECIALLY IN RELATION TO SHOCK AND VISCERAL LESIONS   Delivered at the Royal College of Surgeons. 8vo. cloth, 10s. 6d.

II.

OUTLINES OF SURGERY ; being an Epitome of the Lectures on the Principles and the Practice of Surgery delivered at St. Thomas's Hospital. Fcap. 8vo. cloth, 5s.

JOHN CLAY, M.R.C.S.

KIWISCH ON DISEASES OF THE OVARIES: Translated, by permission, from the last German Edition of his Clinical Lectures on the Special Pathology and Treatment of the Diseases of Women. With Notes, and an Appendix on the Operation of Ovariotomy. Royal 12mo. cloth, 16s.

OAKLEY COLES.

DEFORMITIES OF THE MOUTH; CONGENITAL and ACQUIRED ; their Mechanical Treatment. With Coloured Plates. Second Edition, 8vo., 5s. 6d.

MAURICE H. COLLIS, M.D.DUB., F.R.C.S.I.

THE DIAGNOSIS AND TREATMENT OF CANCER AND THE TUMOURS ANALOGOUS TO IT. With coloured Plates. 8vo. cloth, 14s.

A. J. COOLEY.

THE CYCLOPÆDIA OF PRACTICAL RECEIPTS, PROCESSES, AND COLLATERAL INFORMATION IN THE ARTS, MANUFACTURES, PROFESSIONS, AND TRADES, INCLUDING MEDICINE, PHARMACY, AND DOMESTIC ECONOMY; designed as a General Book of Reference for the Manufacturer, Tradesman, Amateur, and Heads of Families. Fourth and greatly enlarged Edition, 8vo. cloth, 28s.

W. WHITE COOPER, F.R.C.S.

I.

ON WOUNDS AND INJURIES OF THE EYE. Illustrated by 17 Coloured Figures and 41 Woodcuts. 8vo. cloth, 12s.

II.

ON NEAR SIGHT, AGED SIGHT, IMPAIRED VISION, AND THE MEANS OF ASSISTING SIGHT. With 31 Illustrations on Wood. Second Edition. Fcap. 8vo. cloth, 7s. 6d.

S. COOPER.

A DICTIONARY OF PRACTICAL SURGERY AND ENCYCLO-
PÆDIA OF SURGICAL SCIENCE. New Edition, brought down to the present
time. By SAMUEL A. LANE, F.R.C.S., assisted by various eminent Surgeons. Vol. I.,
8vo. cloth, £1. 5s.

HOLMES COOTE, F.R.C.S.

A REPORT ON SOME IMPORTANT POINTS IN THE
TREATMENT OF SYPHILIS. 8vo. cloth, 5s.

R. P. COTTON, M.D., F.R.C.P.

PHTHISIS AND THE STETHOSCOPE; OR, THE PHYSICAL
SIGNS OF CONSUMPTION. Fourth Edition. Foolscap 8vo. cloth, 3s. 6d.

WALTER J. COULSON, F.R.C.S.

I.

A TREATISE ON SYPHILIS. 8vo. cloth, 10s.

II.

STONE IN THE BLADDER: Its Prevention, Early Symptoms, and
Treatment by Lithotrity. 8vo. cloth, 6s.

T. B. CURLING, F.R.C.S., F.R.S.

I.

OBSERVATIONS ON DISEASES OF THE RECTUM. Third
Edition. 8vo. cloth, 7s. 6d.   II.

A PRACTICAL TREATISE ON DISEASES OF THE TESTIS,
SPERMATIC CORD, AND SCROTUM. Third Edition, with Engravings. 8vo.
cloth, 16s.

WILLIAM DALE, M.D.LOND.

A COMPENDIUM OF PRACTICAL MEDICINE AND MORBID
ANATOMY. With Plates, 12mo. cloth, 7s.

DONALD DALRYMPLE, M.P., M.R.C.P.

THE CLIMATE OF EGYPT: METEOROLOGICAL AND MEDI-
CAL OBSERVATIONS, with Practical Hints for Invalid Travellers. Post 8vo. cloth, 4s.

JOHN DALRYMPLE, F.R.C.S., F.R.S.

PATHOLOGY OF THE HUMAN EYE. Complete in Nine Fasciculi:
imperial 4to., 20s. each; half-bound morocco, gilt tops, 9l. 15s.

FRED. DAVIES, M.D., F.R.C.S.

THE UNITY OF MEDICINE: its CORRUPTIONS and DIVI-
SIONS by LAW ESTABLISHED; their Causes, Effects and Remedy. With a
Coloured Chart. Second Edition. 8vo., 10s.

HERBERT DAVIES, M.D., F.R.C.P.

ON THE PHYSICAL DIAGNOSIS OF DISEASES OF THE
LUNGS AND HEART. Second Edition. Post 8vo. cloth, 8s.

JAMES G. DAVEY, M.D., M.R.C.P.

I.

THE GANGLIONIC NERVOUS SYSTEM: its Structure, Functions,
and Diseases. 8vo. cloth, 9s.   II.

ON THE NATURE AND PROXIMATE CAUSE OF IN-
SANITY. Post 8vo. cloth, 3s.

HENRY DAY, M.D., M.R.C.P.

CLINICAL HISTORIES; with Comments. 8vo. cloth, 7s. 6d.

JAMES DIXON, F.R.C.S.

A GUIDE TO THE PRACTICAL STUDY OF DISEASES OF THE EYE. Third Edition. Post 8vo. cloth, 9s.

HORACE DOBELL, M.D.

I.

DEMONSTRATIONS OF DISEASES IN THE CHEST, AND THEIR PHYSICAL DIAGNOSIS. With Coloured Plates. 8vo. cloth, 12s. 6d.

II.

LECTURES ON THE GERMS AND VESTIGES OF DISEASE, and on the Prevention of the Invasion and Fatality of Disease by Periodical Examinations. 8vo. cloth, 6s. 6d.

III.

ON TUBERCULOSIS: ITS NATURE, CAUSE, AND TREAT-MENT; with Notes on Pancreatic Juice. Second Edition. Crown 8vo. cloth, 3s. 6d.

IV.

LECTURES ON WINTER COUGH (CATARRH, BRONCHITIS, EMPHYSEMA, ASTHMA); with an Appendix on some Principles of Diet in Disease. Post 8vo. cloth, 5s. 6d.

V.

LECTURES ON THE TRUE FIRST STAGE OF CONSUMP-TION. Crown 8vo. cloth, 3s. 6d.

O. TOOGOOD DOWNING, M.D.

NEURALGIA: its various Forms, Pathology, and Treatment. THE JACKSONIAN PRIZE ESSAY FOR 1850. 8vo. cloth, 10s. 6d.

ROBERT DRUITT, F.R.C.S.

THE SURGEON'S VADE-MECUM; with numerous Engravings on Wood. Tenth Edition. Foolscap 8vo. cloth, 12s. 6d.

ERNEST EDWARDS, B.A.

PHOTOGRAPHS OF EMINENT MEDICAL MEN, with brief Analytical Notices of their Works. Vols. I. and II. (24 Portraits), 4to. cloth, 24s. each.

CHARLES ELAM, M.D., F.R.C.P.

MEDICINE, DISEASE, AND DEATH: being an Enquiry into the Progress of Medicine as a Practical Art. 8vo. cloth, 3s. 6d.

EDWARD ELLIS, M.D.

A PRACTICAL MANUAL OF THE DISEASES OF CHILDREN. With a Formulary. Crown 8vo. cloth, 6s.

SIR JAMES EYRE, M.D.

I.

THE STOMACH AND ITS DIFFICULTIES. Sixth Edition, by Mr. BEALE. Fcap. 8vo., 2s. 6d.

II.

PRACTICAL REMARKS ON SOME EXHAUSTING DIS-EASES. Second Edition. Post 8vo. cloth, 4s. 6d.

J. FAYRER, M.D., F.R.C.S., C.S.I.

CLINICAL SURGERY IN INDIA. With Engravings. 8vo. cloth, 16s.

SAMUEL FENWICK, M.D., F.R.C.P.

I.

THE MORBID STATES OF THE STOMACH AND DUO-
DENUM, AND THEIR RELATIONS TO THE DISEASES OF OTHER
ORGANS. With 10 Plates. 8vo. cloth, 12s.

II.

THE STUDENT'S GUIDE TO MEDICAL DIAGNOSIS. With
41 Engravings. Fcap. 8vo. cloth, 5s. 6d.

SIR WILLIAM FERGUSSON, BART., F.R.C.S., F.R.S.

I.

A SYSTEM OF PRACTICAL SURGERY; with numerous Illustrations on Wood. Fourth Edition. Fcap. 8vo. cloth, 12s. 6d.

II.

LECTURES ON THE PROGRESS OF ANATOMY AND
SURGERY DURING THE PRESENT CENTURY. With numerous Engravings.
8vo. cloth, 10s. 6d.

SIR JOHN FIFE, F.R.C.S. AND DAVID URQUHART.

MANUAL OF THE TURKISH BATH. Heat a Mode of Cure and
a Source of Strength for Men and Animals. With Engravings. Post 8vo. cloth, 5s.

W. H. FLOWER, F.R.C.S., F.R.S.

DIAGRAMS OF THE NERVES OF THE HUMAN BODY,
exhibiting their Origin, Divisions, and Connexions, with their Distribution to the various
Regions of the Cutaneous Surface, and to all the Muscles. Folio, containing Six
Plates, 14s.

WILLIAM FLUX.

THE LAW TO REGULATE THE SALE OF POISONS WITHIN
GREAT BRITAIN. Crown 8vo. cloth, 2s. 6d.

G. FOWNES, PH.D., F.R.S.

I.

A MANUAL OF CHEMISTRY; with 187 Illustrations on Wood.
Tenth Edition. Fcap. 8vo. cloth, 14s.
Edited by H. BENCE JONES, M.D., F.R.S., and HENRY WATTS, B.A., F.R.S.

II.

CHEMISTRY, AS EXEMPLIFYING THE WISDOM AND
BENEFICENCE OF GOD. Second Edition. Fcap. 8vo. cloth, 4s. 6d.

III.

INTRODUCTION TO QUALITATIVE ANALYSIS. Post 8vo. cloth, 2s.

D. J. T. FRANCIS, M.D., F.R.C.P.

CHANGE OF CLIMATE; considered as a Remedy in Dyspeptic, Pul-
monary, and other Chronic Affections; with an Account of the most Eligible Places of
Residence for Invalids, at different Seasons of the Year. Post 8vo. cloth, 8s. 6d.

W. H. FULLER, M.D., F.R.C.P.

I.

ON DISEASES OF THE LUNGS AND AIR PASSAGES.
Second Edition. 8vo. cloth, 12s. 6d.

II.

ON DISEASES OF THE HEART AND GREAT VESSELS.
8vo. cloth, 7s. 6d.

III.

ON RHEUMATISM, RHEUMATIC GOUT, AND SCIATICA:
their Pathology, Symptoms, and Treatment. Third Edition. 8vo. cloth, 12s. 6d.

REMIGIUS FRESENIUS.

A SYSTEM OF INSTRUCTION IN CHEMICAL ANALYSIS,
Edited by ARTHUR VACHER.
QUALITATIVE. Seventh Edition. 8vo. cloth, 9s.
QUANTITATIVE. Fifth Edition. 8vo. cloth, 12s. 6d.

ROBERT GALLOWAY.

I.

THE FIRST STEP IN CHEMISTRY. With numerous Engravings.
Fourth Edition. Fcap. 8vo. cloth, 6s. 6d.

II.

A KEY TO THE EXERCISES CONTAINED IN ABOVE. Fcap.
8vo., 2s. 6d.

III.

THE SECOND STEP IN CHEMISTRY; or, the Student's Guide to
the Higher Branches of the Science. With Engravings. 8vo. cloth, 10s.

IV.

A MANUAL OF QUALITATIVE ANALYSIS. Fifth Edition.
With Engravings. Post 8vo. cloth, 8s. 6d.

V.

CHEMICAL TABLES. On Five Large Sheets, for School and Lecture
Rooms. Second Edition. 4s. 6d.

J. SAMPSON GAMGEE, M.R.C.S.

HISTORY OF A SUCCESSFUL CASE OF AMPUTATION AT
THE HIP-JOINT (the limb 48-in. in circumference, 99 pounds weight). With 4
Photographs. 4to cloth, 10s. 6d.

F. J. GANT, F.R.C.S.

I.

THE PRINCIPLES OF SURGERY: Clinical, Medical, and Operative. With Engravings. 8vo. cloth, 18s.

II.

THE IRRITABLE BLADDER: its Causes and Curative Treatment.
Second Edition, enlarged. Crown 8vo. cloth, 5s.

C. B. GARRETT, M.D.

IRRITATIVE DYSPEPSIA AND ITS IMPORTANT CON-
NECTION with IRRITATIVE CONGESTION of the WINDPIPE, and with
the Origin and Progress of Consumption. Crown 8vo., 2s. 6d.

JOHN GAY, F.R.C.S.

ON VARICOSE DISEASE OF THE LOWER EXTREMITIES.
LETTSOMIAN LECTURES. With Plates. 8vo. cloth, 5s.

SIR DUNCAN GIBB, BART., M.D.

I.

ON DISEASES OF THE THROAT AND WINDPIPE, as
reflected by the Laryngoscope. Second Edition. With 116 Engravings. Post 8vo.
cloth, 10s. 6d.

II.

THE LARYNGOSCOPE IN DISEASES OF THE THROAT.
with a Chapter on RHINOSCOPY. Third Edition, with Engravings. Crown 8vo.,
cloth, 5s.

C. A. GORDON M.D., C.B.

I.,

ARMY HYGIENE. 8vo. cloth, 20s.

II.

CHINA, FROM A MEDICAL POINT OF VIEW; IN 1860 AND 1861; With a Chapter on Nagasaki as a Sanatarium. 8vo. cloth, 10s. 6d.

WILLIAM GAIRDNER, M.D.

ON GOUT; its History, its Causes, and its Cure. Fourth Edition. Post 8vo. cloth, 8s. 6d.

MICHAEL C. GRABHAM, M.D., M.R.C.P.

THE CLIMATE AND RESOURCES OF MADEIRA, as regarding chiefly the Necessities of Consumption and the Welfare of Invalids. With Map and Engravings. Crown 8vo. cloth, 5s.

R. J. GRAVES, M.D., F.R.S.

STUDIES IN PHYSIOLOGY AND MEDICINE. Edited by Dr. Stokes. With Portrait and Memoir. 8vo. cloth, 14s.

T. GRIFFITHS.

CHEMISTRY OF THE FOUR SEASONS — Spring, Summer, Autumn, Winter. Illustrated with Engravings on Wood. Second Edition. Foolscap 8vo. cloth, 7s. 6d.

JAMES M. GULLY, M.D.

THE SIMPLE TREATMENT OF DISEASE; deduced from the Methods of Expectancy and Revulsion. 18mo. cloth, 4s.

W. A. GUY, M.B., F.R.S., AND JOHN HARLEY, M.D., F.R.C.P.

HOOPER'S PHYSICIAN'S VADE-MECUM: OR, MANUAL OF THE PRINCIPLES AND PRACTICE OF PHYSIC. Seventh Edition. With Engravings. Foolscap 8vo. cloth, 12s. 6d.

GUY'S HOSPITAL REPORTS. Third Series. Vol. XV., 8vo. 7s. 6d.

S. O. HABERSHON, M.D., F.R.C.P.

I.

ON DISEASES OF THE ABDOMEN, comprising those of the Stomach and other Parts of the Alimentary Canal, Œsophagus, Stomach, Cæcum, Intestines, and Peritoneum. Second Edition, with Plates. 8vo. cloth, 14s.

II.

ON THE INJURIOUS EFFECTS OF MERCURY IN THE TREATMENT OF DISEASE. Post 8vo. cloth, 3s. 6d.

C. RADCLYFFE HALL, F.R.C.P.

TORQUAY IN ITS MEDICAL ASPECT AS A RESORT FOR PULMONARY INVALIDS. Post 8vo. cloth, 5s.

MARSHALL HALL, M.D., F.R.S.

I.

PRONE AND POSTURAL RESPIRATION IN DROWNING AND OTHER FORMS OF APNŒA OR SUSPENDED RESPIRATION. Post 8vo. cloth. 5s.

II.

PRACTICAL OBSERVATIONS AND SUGGESTIONS IN MEDI-CINE. Second Series. Post 8vo. cloth, 8s. 6d.

REV. T. F. HARDWICH.

A MANUAL OF PHOTOGRAPHIC CHEMISTRY. With Engravings. Seventh Edition. Foolscap 8vo. cloth, 7s. 6d.

J. BOWER HARRISON, M.D., M.R.C.P.

I.

LETTERS TO A YOUNG PRACTITIONER ON THE DISEASES OF CHILDREN. Foolscap 8vo. cloth, 3s.

II.

ON THE CONTAMINATION OF WATER BY THE POISON OF LEAD, and its Effects on the Human Body. Foolscap 8vo. cloth, 3s. 6d.

GEORGE HARTWIG, M.D.

I.

ON SEA BATHING AND SEA AIR. Second Edition. Fcap. 8vo., 2s. 6d.

II.

ON THE PHYSICAL EDUCATION OF CHILDREN. Fcap. 8vo., 2s. 6d.

A. H. HASSALL, M.D.

THE URINE, IN HEALTH AND DISEASE; being an Explanation of the Composition of the Urine, and of the Pathology and Treatment of Urinary and Renal Disorders. Second Edition. With 79 Engravings (23 Coloured). Post 8vo. cloth, 12s. 6d.

ALFRED HAVILAND, M.R.C.S.

CLIMATE, WEATHER, AND DISEASE; being a Sketch of the Opinions of the most celebrated Ancient and Modern Writers with regard to the Influence of Climate and Weather in producing Disease. With Four coloured Engravings. 8vo. cloth, 7s.

W. HAYCOCK, M.R.C.V.S.

HORSES; HOW THEY OUGHT TO BE SHOD: being a plain and practical Treatise on the Principles and Practice of the Farrier's Art. With 14 Plates. Cloth, 7s. 6d.

F. W. HEADLAND, M.D., F.R.C.P.

I.

ON THE ACTION OF MEDICINES IN THE SYSTEM. Fourth Edition. 8vo. cloth, 14s.

II.

A MEDICAL HANDBOOK; comprehending such Information on Medical and Sanitary Subjects as is desirable in Educated Persons. Second Thousand. Foolscap 8vo. cloth, 5s.

J. N. HEALE, M.D., M.R.C.P.

I.

A TREATISE ON THE PHYSIOLOGICAL ANATOMY OF THE LUNGS. With Engravings. 8vo. cloth, 8s.

II.

A TREATISE ON VITAL CAUSES. 8vo. cloth, 9s.

CHRISTOPHER HEATH, F.R.C.S.

I.

PRACTICAL ANATOMY: a Manual of Dissections. With numerous Engravings. Second Edition. Fcap. 8vo. cloth, 12s. 6d.

II.

A MANUAL OF MINOR SURGERY AND BANDAGING, FOR THE USE OF HOUSE-SURGEONS, DRESSERS, AND JUNIOR PRACTITIONERS. With Illustrations. Fourth Edition. Fcap. 8vo. cloth, 5s. 6d.

III.

INJURIES AND DISEASES OF THE JAWS. JACKSONIAN PRIZE ESSAY. With Engravings. 8vo. cloth, 12s.

JOHN HIGGINBOTTOM, F.R.S., F.R.C.S.E.

A PRACTICAL ESSAY ON THE USE OF THE NITRATE OF SILVER IN THE TREATMENT OF INFLAMMATION, WOUNDS, AND ULCERS. Third Edition, 8vo. cloth, 6s.

WILLIAM HINDS, M.D.

THE HARMONIES OF PHYSICAL SCIENCE IN RELATION TO THE HIGHER SENTIMENTS; with Observations on Medical Studies, and on the Moral and Scientific Relations of Medical Life. Post 8vo. cloth, 4s.

J. A. HINGESTON, M.R.C.S.

TOPICS OF THE DAY, MEDICAL, SOCIAL, AND SCIENTIFIC. Crown 8vo. cloth, 7s. 6d.

RICHARD HODGES, M.D.

THE NATURE, PATHOLOGY, AND TREATMENT OF PUERPERAL CONVULSIONS. Crown 8vo. cloth, 3s.

DECIMUS HODGSON, M.D.

THE PROSTATE GLAND, AND ITS ENLARGEMENT IN OLD AGE. With 12 Plates. Royal 8vo. cloth, 6s.

JABEZ HOGG, M.R.C.S.

A MANUAL OF OPHTHALMOSCOPIC SURGERY; being a Practical Treatise on the Use of the Ophthalmoscope in Diseases of the Eye. Third Edition. With Coloured Plates. 8vo. cloth, 10s. 6d.

LUTHER HOLDEN, F.R.C.S.

I.

HUMAN OSTEOLOGY: with Plates, showing the Attachments of the Muscles. Fourth Edition. 8vo. cloth, 16s.

II.

A MANUAL OF THE DISSECTION OF THE HUMAN BODY. With Engravings on Wood. Third Edition. 8vo. cloth, 16s.

BARNARD HOLT, F.R.C.S.

ON THE IMMEDIATE TREATMENT OF STRICTURE OF THE URETHRA. Third Edition, Enlarged. 8vo. cloth, 6s.

C. HOLTHOUSE, F.R.C.S.

ON HERNIAL AND OTHER TUMOURS OF THE GROIN

and its NEIGHBOURHOOD with some Practical Remarks on the Radical Cure of Ruptures. 8vo., 6s. 6d.

P. HOOD M.D.

THE SUCCESSFUL TREATMENT OF SCARLET FEVER;

also, OBSERVATIONS ON THE PATHOLOGY AND TREATMENT OF CROWING INSPIRATIONS OF INFANTS. Post 8vo. cloth, 5s.

JOHN HORSLEY.

A CATECHISM OF CHEMICAL PHILOSOPHY; being a Familiar

Exposition of the Principles of Chemistry and Physics. With Engravings on Wood. Designed for the Use of Schools and Private Teachers. Post 8vo. cloth, 6s. 6d.

JAMES A. HORTON, M.D.

PHYSICAL AND MEDICAL CLIMATE AND METEOROLOGY

OF THE WEST COAST OF AFRICA. 8vo. cloth, 10s.

LUKE HOWARD, F.R.S.

ESSAY ON THE MODIFICATIONS OF CLOUDS. Third Edition,

by W. D. and E. Howard. With 6 Lithographic Plates, from Pictures by Kenyon. 4to. cloth, 10s. 6d.

A. HAMILTON HOWE, M.D.

A THEORETICAL INQUIRY INTO THE PHYSICAL CAUSE

OF EPIDEMIC DISEASES. Accompanied with Tables. 8vo. cloth, 7s.

C. W. HUFELAND.

THE ART OF PROLONGING LIFE. Second Edition. Edited

by Erasmus Wilson, F.R.S. Foolscap 8vo., 2s. 6d.

W. CURTIS HUGMAN, F.R.C.S.

ON HIP-JOINT DISEASE; with reference especially to Treatment

by Mechanical Means for the Relief of Contraction and Deformity of the Affected Limb. With Plates. Re-issue, enlarged. 8vo. cloth, 3s. 6d.

J. W. HULKE, F.R.C.S., F.R.S.

A PRACTICAL TREATISE ON THE USE OF THE

OPHTHALMOSCOPE. Being the Jacksonian Prize Essay for 1859. Royal 8vo. cloth, 8s.

HENRY HUNT, F.R.C.P.

ON HEARTBURN AND INDIGESTION. 8vo. cloth, 5s.

G. Y. HUNTER, M.R.C.S.

BODY AND MIND: the Nervous System and its Derangements.

Fcap. 8vo. cloth, 3s. 6d.

JONATHAN HUTCHINSON, F.R.C.S.

A CLINICAL MEMOIR ON CERTAIN DISEASES OF THE
EYE AND EAR, CONSEQUENT ON INHERITED SYPHILIS; with an
appended Chapter of Commentaries on the Transmission of Syphilis from Parent to
Offspring, and its more remote Consequences.   With Plates and Woodcuts, 8vo. cloth, 9s.

T. H. HUXLEY, LL.D., F.R.S.

INTRODUCTION TO THE CLASSIFICATION OF ANIMALS.
With Engravings.   8vo. cloth, 6s.

THOMAS INMAN, M.D., M.R.C.P.
I.

ON MYALGIA: ITS NATURE, CAUSES, AND TREATMENT;
being a Treatise on Painful and other Affections of the Muscular System.   Second
Edition.   8vo. cloth, 9s.

II.

FOUNDATION FOR A NEW THEORY AND PRACTICE
OF MEDICINE.   Second Edition.   Crown 8vo. cloth, 10s.

JAMES JAGO, M.D.OXON., A.B.CANTAB.

ENTOPTICS, WITH ITS USES IN PHYSIOLOGY AND
MEDICINE.   With 54 Engravings.   Crown 8vo. cloth, 5s.

M. PROSSER JAMES, M.D., M.R.C.P.

SORE-THROAT: ITS NATURE, VARIETIES, AND TREAT-
MENT; including the Use of the LARYNGOSCOPE as an Aid to Diagnosis.   Second
Edition, with numerous Engravings.   Post 8vo. cloth, 5s.

F. E. JENCKEN, M.D., M.R.C.P.

THE CHOLERA: ITS ORIGIN, IDIOSYNCRACY, AND
TREATMENT.   Fcap. 8vo. cloth, 2s. 6d.

C. HANDFIELD JONES, M.B., F.R.C.P., F.R.S.

STUDIES ON FUNCTIONAL NERVOUS DISORDERS.   Second
Edition, much enlarged.   8vo. cloth, 18s.

H. BENCE JONES, M.D., F.R.C.P., D.C.L., F.R.S.
I.

LECTURES ON SOME OF THE APPLICATIONS OF
CHEMISTRY AND MECHANICS TO PATHOLOGY AND THERA-
PEUTICS.   8vo. cloth, 12s.

II.

CROONIAN LECTURES ON MATTER AND FORCE.   Fcap. 8vo.
cloth, 5s.

C. HANDFIELD JONES, M.B., F.R.S., & E. H. SIEVEKING, M.D., F.R.C.P.

A MANUAL OF PATHOLOGICAL ANATOMY.   Illustrated with
numerous Engravings on Wood.   Foolscap 8vo. cloth, 12s. 6d.

JAMES JONES, M.D., M.R.C.P.

ON THE USE OF PERCHLORIDE OF IRON AND OTHER
CHALYBEATE SALTS IN THE TREATMENT OF CONSUMPTION. Crown 8vo. cloth, 3s. 6d.

T. WHARTON JONES, F.R.C.S., F.R.S.

I.

A MANUAL OF THE PRINCIPLES AND PRACTICE OF
OPHTHALMIC MEDICINE AND SURGERY; with Nine Coloured Plates and 173 Wood Engravings. Third Edition, thoroughly revised. Foolscap 8vo. cloth, 12s. 6d.

II.

THE WISDOM AND BENEFICENCE OF THE ALMIGHTY,
AS DISPLAYED IN THE SENSE OF VISION. Actonian Prize Essay. With Illustrations on Steel and Wood. Foolscap 8vo. cloth, 4s. 6d.

III.

DEFECTS OF SIGHT AND HEARING: their Nature, Causes, Prevention, and General Management. Second Edition, with Engravings. Fcap. 8vo. 2s. 6d.

IV.

A CATECHISM OF THE MEDICINE AND SURGERY OF
THE EYE AND EAR. For the Clinical Use of Hospital Students. Fcap. 8vo. 2s. 6d.

V.

A CATECHISM OF THE PHYSIOLOGY AND PHILOSOPHY
OF BODY, SENSE, AND MIND. For Use in Schools and Colleges. Fcap. 8vo., 2s. 6d.

U. J. KAY-SHUTTLEWORTH, M.P.

FIRST PRINCIPLES OF MODERN CHEMISTRY: a Manual
of Inorganic Chemistry. Second Edition. Crown 8vo. cloth, 4s. 6d.

DR. LAENNEC.

A MANUAL OF AUSCULTATION AND PERCUSSION. Translated and Edited by J. B. SHARPE, M.R.C.S. 3s.

SIR WM. LAWRENCE, BART., F.R.S.

I.

LECTURES ON SURGERY. 8vo. cloth, 16s.

II.

A TREATISE ON RUPTURES. The Fifth Edition, considerably
enlarged. 8vo. cloth, 16s.

ARTHUR LEARED, M.D., M.R.C.P.

IMPERFECT DIGESTION: ITS CAUSES AND TREATMENT.
Fifth Edition. Foolscap 8vo. cloth, 4s. 6d.

HENRY LEE, F.R.C.S.

PRACTICAL PATHOLOGY. Third Edition, in 2 Vols. Containing
Lectures on Suppurative Fever, Diseases of the Veins, Hæmorrhoidal Tumours, Diseases of the Rectum, Syphilis, Gonorrhœal Ophthalmia, &c. 8vo. cloth, 10s. each vol.

EDWIN LEE, M.D.

I.

THE EFFECT OF CLIMATE ON TUBERCULOUS DISEASE,
with Notices of the chief Foreign Places of Winter Resort. Small 8vo. cloth, 4s. 6d.

II.

THE WATERING PLACES OF ENGLAND, CONSIDERED
with Reference to their Medical Topography. Fourth Edition. Fcap. 8vo. cloth, 7s. 6d.

III.

THE BATHS OF FRANCE. Fourth Edition. Fcap. 8vo. cloth,
4s. 6d.

IV.

THE BATHS OF GERMANY. Fourth Edition. Post 8vo. cloth, 7s.

V.

THE BATHS OF SWITZERLAND. 12mo. cloth, 3s. 6d.

VI.

HOMŒOPATHY AND HYDROPATHY IMPARTIALLY AP-
PRECIATED. Fourth Edition. Post 8vo. cloth, 3s.

ROBERT LEE, M.D, F.R.C.P., F.R.S.

I.

CONSULTATIONS IN MIDWIFERY. Foolscap 8vo. cloth, 4s. 6d.

II.

A TREATISE ON THE SPECULUM; with Three Hundred Cases.
8vo. cloth, 4s. 6d.

III.

CLINICAL REPORTS OF OVARIAN AND UTERINE DIS-
EASES, with Commentaries. Foolscap 8vo. cloth, 6s. 6d.

IV.

CLINICAL MIDWIFERY: comprising the Histories of 545 Cases of
Difficult, Preternatural, and Complicated Labour, with Commentaries. Second Edition. Foolscap 8vo. cloth, 5s.

WM. LEISHMAN, M.D., F.F.P.S.

THE MECHANISM OF PARTURITION: An Essay, Historical and
Critical. With Engravings. 8vo. cloth, 5s.

F. HARWOOD LESCHER.

THE ELEMENTS OF PHARMACY. 8vo. cloth, 7s. 6d.

ROBERT LISTON, F.R.S.

PRACTICAL SURGERY. Fourth Edition. 8vo. cloth, 22s.

D. D. LOGAN, M.D., M.R.C.P. LOND.

ON OBSTINATE DISEASES OF THE SKIN. Fcap. 8vo. cloth, 2s. 6d.

LONDON HOSPITAL.

CLINICAL LECTURES AND REPORTS BY THE MEDICAL
AND SURGICAL STAFF. With Illustrations. Vols. I. to IV. 8vo. cloth, 7s. 6d.

LONDON MEDICAL SOCIETY OF OBSERVATION.

WHAT TO OBSERVE AT THE BED-SIDE, AND AFTER DEATH. Published by Authority. Second Edition. Foolscap 8vo. cloth, 4s. 6d.

HENRY LOWNDES, M.R.C.S.

AN ESSAY ON THE MAINTENANCE OF HEALTH. Fcap. 8vo. cloth, 2s. 6d.

MORELL MACKENZIE, M.D. LOND., M.R.C.P.

HOARSENESS, LOSS OF VOICE, AND STRIDULOUS BREATHING in relation to NERVO-MUSCULAR AFFECTIONS of the LARYNX. Second Edition. Fully Illustrated. 8vo. 2s. 6d.

DANIEL MACLACHLAN, M.D., F.R.C.P.L.

THE DISEASES AND INFIRMITIES OF ADVANCED LIFE. 8vo. cloth, 16s.

A. C. MACLEOD, M.R.C.P.LOND.

ACHOLIC DISEASES; comprising Jaundice, Diarrhœa, Dysentery, and Cholera. Post 8vo. cloth, 5s. 6d.

GEORGE H. B. MACLEOD, M.D., F.R.C.S.EDIN.

I.

OUTLINES OF SURGICAL DIAGNOSIS. 8vo. cloth, 12s. 6d.

II.

NOTES ON THE SURGERY OF THE CRIMEAN WAR; with REMARKS on GUN-SHOT WOUNDS. 8vo. cloth, 10s. 6d.

WM. MACLEOD, M.D., F.R.C.P.EDIN.

THE THEORY OF THE TREATMENT OF DISEASE ADOPTED AT BEN RHYDDING. Fcap. 8vo. cloth, 2s. 6d.

JOSEPH MACLISE, F.R.C.S.

I.

SURGICAL ANATOMY. A Series of Dissections, illustrating the Principal Regions of the Human Body. Second Edition, folio, cloth, £3. 12s.; half-morocco, £4. 4s.

II.

ON DISLOCATIONS AND FRACTURES. This Work is Uniform with "Surgical Anatomy;" folio, cloth, £2. 10s.; half-morocco, £2. 17s.

N. C. MACNAMARA.

I.

A MANUAL OF THE DISEASES OF THE EYE. With Coloured Plates. Fcap. 8vo. cloth, 12s. 6d.

II.

A TREATISE ON ASIATIC CHOLERA; with Maps. 8vo. cloth, 16s.

WM. MARCET, M.D., F.R.C.P., F.R.S.

ON CHRONIC ALCOHOLIC INTOXICATION; with an INQUIRY INTO THE INFLUENCE OF THE ABUSE OF ALCOHOL AS A PRE-DISPOSING CAUSE OF DISEASE. Second Edition, much enlarged. Foolscap 8vo. cloth, 4s. 6d.

J. MACPHERSON, M.D.

CHOLERA IN ITS HOME; with a Sketch of the Pathology and Treatment of the Disease. Crown 8vo. cloth, 5s.

W. O. MARKHAM, M.D., F.R.C.P.

I.

DISEASES OF THE HEART: THEIR PATHOLOGY, DIAGNOSIS, AND TREATMENT. Second Edition. Post 8vo. cloth, 6s.

II.

SKODA ON AUSCULTATION AND PERCUSSION. Post 8vo. cloth, 6s.

III.

BLEEDING AND CHANGE IN TYPE OF DISEASES. Gulstonian Lectures for 1864. Crown 8vo. 2s. 6d.

ALEXANDER MARSDEN, M.D., F.R.C.S.

A NEW AND SUCCESSFUL MODE OF TREATING CERTAIN FORMS OF CANCER; to which is prefixed a Practical and Systematic Description of all the Varieties of this Disease. With Coloured Plates. 8vo. cloth, 6s. 6d.

SIR RANALD MARTIN, C.B., F.R.C.S., F.R.S.

INFLUENCE OF TROPICAL CLIMATES IN PRODUCING THE ACUTE ENDEMIC DISEASES OF EUROPEANS; including Practical Observations on their Chronic Sequelæ under the Influences of the Climate of Europe. Second Edition, much enlarged. 8vo. cloth, 20s.

P. MARTYN, M.D.LOND.

HOOPING-COUGH; ITS PATHOLOGY AND TREATMENT. With Engravings. 8vo. cloth, 2s. 6d.

C. F. MAUNDER, F.R.C.S.

OPERATIVE SURGERY. With 158 Engravings. Post 8vo. 6s.

R. G. MAYNE, M.D., LL.D.

I.

AN EXPOSITORY LEXICON OF THE TERMS, ANCIENT AND MODERN, IN MEDICAL AND GENERAL SCIENCE. 8vo. cloth, £2. 10s.

II.

A MEDICAL VOCABULARY; or, an Explanation of all Names, Synonymes, Terms, and Phrases used in Medicine and the relative branches of Medical Science. Third Edition. Fcap. 8vo. cloth, 8s. 6d.

EDWARD MERYON, M.D., F.R.C.P.

PATHOLOGICAL AND PRACTICAL RESEARCHES ON THE VARIOUS FORMS OF PARALYSIS. 8vo. cloth, 6s.

W. J. MOORE, M.D.

I.

HEALTH IN THE TROPICS; or, Sanitary Art applied to Europeans in India. 8vo. cloth, 9s.

II.

A MANUAL OF THE DISEASES OF INDIA. Fcap. 8vo. cloth, 5s.

JAMES MORRIS, M.D.LOND.

I.

GERMINAL MATTER AND THE CONTACT THEORY: An Essay on the Morbid Poisons. Second Edition. Crown 8vo. cloth, 4s. 6d.

II.

IRRITABILITY: Popular and Practical Sketches of Common Morbid States and Conditions bordering on Disease; with Hints for Management, Alleviation, and Cure. Crown 8vo. cloth, 4s. 6d.

G. J. MULDER.

THE CHEMISTRY OF WINE. Edited by H. BENCE JONES, M.D., F.R.S. Fcap. 8vo. cloth, 6s.

W. MURRAY, M.D., M.R.C.P.

EMOTIONAL DISORDERS OF THE SYMPATHETIC SYSTEM OF NERVES. Crown 8vo. cloth, 3s. 6d.

W. B. MUSHET, M.B., M.R.C.P.

ON APOPLEXY, AND ALLIED AFFECTIONS OF THE BRAIN. 8vo. cloth, 7s.

ARTHUR B. R. MYERS, M.R.C.S.

ON THE ETIOLOGY AND PREVALENCE OF DISEASES of the HEART among SOLDIERS. With Diagrams. The "Alexander" Prize Essay. 8vo., 4s.

GEORGE NAYLER, F.R.C.S.

ON THE DISEASES OF THE SKIN. With Plates. 8vo. cloth, 10s. 6d.

J. BIRKBECK NEVINS, M.D.

THE PRESCRIBER'S ANALYSIS OF THE BRITISH PHARMACOPEIA of 1867. 32mo. cloth, 3s. 6d.

H. M. NOAD, PH.D., F.R.S.

THE INDUCTION COIL, being a Popular Explanation of the Electrical Principles on which it is constructed. Third Edition. With Engravings. Fcap. 8vo. cloth, 3s.

DANIEL NOBLE, M.D., F.R.C.P.

THE HUMAN MIND IN ITS RELATIONS WITH THE BRAIN AND NERVOUS SYSTEM. Post 8vo. cloth, 4s. 6d.

SELBY NORTON, M.D.

INFANTILE DISEASES: their Causes, Prevention, and Treatment, showing by what Means the present Mortality may be greatly reduced. Fcap. 8vo. cloth, 2s. 6d.

FRANCIS OPPERT, M.D., M.R.C.P.

I.

HOSPITALS, INFIRMARIES, AND DISPENSARIES; their Construction, Interior Arrangement, and Management, with Descriptions of existing Institutions. With 58 Engravings. Royal 8vo. cloth, 10s. 6d.

II.

VISCERAL AND HEREDITARY SYPHILIS. 8vo. cloth, 5s.

LANGSTON PARKER, F.R.C.S.

THE MODERN TREATMENT OF SYPHILITIC DISEASES, both Primary and Secondary; comprising the Treatment of Constitutional and Confirmed Syphilis, by a safe and successful Method. Fourth Edition, 8vo. cloth, 10s.

E. A. PARKES, M.D., F.R.C.P., F.R.S.

I.

A MANUAL OF PRACTICAL HYGIENE; intended especially for the Medical Officers of the Army. With Plates and Woodcuts. 3rd Edition, 8vo. cloth, 16s.

II.

THE URINE: ITS COMPOSITION IN HEALTH AND DISEASE, AND UNDER THE ACTION OF REMEDIES. 8vo. cloth, 12s.

JOHN PARKIN, M.D., F.R.C.S.

I.

THE ANTIDOTAL TREATMENT AND PREVENTION OF THE EPIDEMIC CHOLERA. Third Edition. 8vo. cloth, 7s. 6d.

II.

THE CAUSATION AND PREVENTION OF DISEASE; with the Laws regulating the Extrication of Malaria from the Surface, and its Diffusion in the surrounding Air. 8vo. cloth, 5s.

JAMES PART, F.R.C.S.

THE MEDICAL AND SURGICAL POCKET CASE BOOK, for the Registration of important Cases in Private Practice, and to assist the Student of Hospital Practice. Second Edition. 2s. 6d.

JOHN PATTERSON, M.D.

EGYPT AND THE NILE AS A WINTER RESORT FOR PULMONARY AND OTHER INVALIDS. Fcap. 8vo. cloth, 3s.

F. W. PAVY, M.D., F.R.S., F.R.C.P.

I.

DIABETES: RESEARCHES ON ITS NATURE AND TREATMENT. Second Edition. With Engravings. 8vo. cloth, 10s.

II.

DIGESTION: ITS DISORDERS AND THEIR TREATMENT. Second Edition. 8vo. cloth, 8s. 6d.

T. B. PEACOCK, M.D., F.R.C.P.

I.

ON MALFORMATIONS OF THE HUMAN HEART. With Original Cases and Illustrations. Second Edition. With 8 Plates. 8vo. cloth, 10s.

II.

ON SOME OF THE CAUSES AND EFFECTS OF VALVULAR DISEASE OF THE HEART. With Engravings. 8vo. cloth, 5s.

W. H. PEARSE, M.D.EDIN.

NOTES ON HEALTH IN CALCUTTA AND BRITISH EMIGRANT SHIPS, including Ventilation, Diet, and Disease. Fcap. 8vo. 2s.

JONATHAN PEREIRA, M.D., F.R.S.

SELECTA E PRÆSCRIPTIS. Fifteenth Edition. 24mo. cloth, 5s.

JAMES H. PICKFORD, M.D.

HYGIENE; or, Health as Depending upon the Conditions of the Atmosphere, Food and Drinks, Motion and Rest, Sleep and Wakefulness, Secretions, Excretions, and Retentions, Mental Emotions, Clothing, Bathing, &c. Vol. I. 8vo. cloth, 9s.

WILLIAM PIRRIE, M.D., C.M:, F.R.S.E.

THE PRINCIPLES AND PRACTICE OF SURGERY. With
numerous Engravings on Wood. Second Edition. 8vo. cloth, 24s.

WILLIAM PIRRIE, M.D.

ON HAY ASTHMA, AND THE AFFECTION TERMED
HAY FEVER. Fcap. 8vo. cloth, 2s. 6d.

HENRY POWER, F.R.C.S., M.B.LOND.

ILLUSTRATIONS OF SOME OF THE PRINCIPAL DISEASES
OF THE EYE : With an Account of their Symptoms, Pathology and Treatment.
Twelve Coloured Plates. 8vo. cloth, 20s.

HENRY F. A. PRATT, M.D., M.R.C.P.

I.

THE GENEALOGY OF CREATION, newly Translated from the
Unpointed Hebrew Text of the Book of Genesis, showing the General Scientific Accuracy
of the Cosmogony of Moses and the Philosophy of Creation. 8vo. cloth, 14s.

II.

ON ECCENTRIC AND CENTRIC FORCE: A New Theory of
Projection. With Engravings. 8vo. cloth, 10s.

III.

ON ORBITAL MOTION: The Outlines of a System of Physical
Astronomy. With Diagrams. 8vo. cloth, 7s. 6d.

IV.

ASTRONOMICAL INVESTIGATIONS. The Cosmical Relations of
the Revolution of the Lunar Apsides. Oceanic Tides. With Engravings. 8vo. cloth, 5s.

V.

THE ORACLES OF GOD: An Attempt at a Re-interpretation. Part I.
The Revealed Cosmos. 8vo. cloth, 10s.

THE PRESCRIBER'S PHARMACOPŒIA; containing all the Medi-
cines in the British Pharmacopœia, arranged in Classes according to their Action, with
their Composition and Doses. By a Practising Physician. Fifth Edition. 32mo.
cloth, 2s. 6d.; roan tuck (for the pocket), 3s. 6d.

JOHN ROWLISON PRETTY, M.D.

AIDS DURING LABOUR, including the Administration of Chloroform,
the Management of Placenta and Post-partum Hæmorrhage. Fcap. 8vo. cloth, 4s. 6d.

P. C. PRICE, F.R.C.S.

AN ESSAY ON EXCISION OF THE KNEE-JOINT. With
Coloured Plates. With Memoir of the Author and Notes by Henry Smith, F.R.C.S.
Royal 8vo. cloth, 14s.

A. E. SANSOM, M.D.LOND., M.R.C.P.

CHLOROFORM: ITS ACTION AND ADMINISTRATION. A Handbook. With Engravings. Crown 8vo. cloth, 5s.

HENRY SAVAGE, M.D.LOND., F.R.C.S.

THE SURGERY, SURGICAL PATHOLOGY, AND SURGICAL ANATOMY of the FEMALE PELVIC ORGANS, in a Series of Coloured Plates taken from Nature. With Commentaries, Notes and Cases. Second Edition, greatly enlarged. 4to., £1. 11s. 6d.

JOHN SAVORY, M.S.A.

A COMPENDIUM OF DOMESTIC MEDICINE, AND COMPANION TO THE MEDICINE CHEST; intended as a Source of Easy Reference for Clergymen, and for Families residing at a Distance from Professional Assistance. Seventh Edition. 12mo. cloth, 5s.

HERMANN SCHACHT.

THE MICROSCOPE, AND ITS APPLICATION TO VEGETABLE ANATOMY AND PHYSIOLOGY. Edited by FREDERICK CURREY, M.A. Post 8vo. cloth, 6s.

R. E. SCORESBY-JACKSON, M.D., F.R.S.E.

MEDICAL CLIMATOLOGY; or, a Topographical and Meteorological Description of the Localities resorted to in Winter and Summer by Invalids of various classes both at Home and Abroad. With an Isothermal Chart. Post 8vo. cloth, 12s.

R. H. SEMPLE M.D., M.R.C.P.

ON COUGH: its Causes, Varieties, and Treatment. With some practical Remarks on the Use of the Stethoscope as an aid to Diagnosis. Post 8vo. cloth, 4s. 6d.

E. J. SEYMOUR, M.D.

I.

ILLUSTRATIONS OF SOME OF THE PRINCIPAL DISEASES OF THE OVARIA: their Symptoms and Treatment; to which are prefixed Observations on the Structure and Functions of those parts in the Human Being and in Animals. On India paper. Folio, 16s.

II.

THE NATURE AND TREATMENT OF DROPSY; considered especially in reference to the Diseases of the Internal Organs of the Body, which most commonly produce it. 8vo. 5s.

THOS. SHAPTER, M.D., F.R.C.P.

THE CLIMATE OF THE SOUTH OF DEVON, AND ITS INFLUENCE UPON HEALTH. Second Edition, with Maps. 8vo. cloth, 10s. 6d.

E. SHAW, M.R.C.S.

THE MEDICAL REMEMBRANCER; OR, BOOK OF EMERGENCIES. Fifth Edition. Edited, with Additions, by JONATHAN HUTCHINSON, F.R.C.S. 32mo. cloth, 2s. 6d.

JOHN SHEA, M.D., B.A.

A MANUAL OF ANIMAL PHYSIOLOGY With an Appendix of Questions for the B.A. London and other Examinations. With Engravings. Foolscap 8vo. cloth, 5s. 6d.

FRANCIS SIBSON, M.D., F.R.C.P., F.R.S.

MEDICAL ANATOMY. With coloured Plates. Imperial folio. Complete in Seven Fasciculi. 5s. each.

E. H. SIEVEKING, M.D., F.R.C.P.

ON EPILEPSY AND EPILEPTIFORM SEIZURES: their Causes, Pathology, and Treatment. Second Edition. Post 8vo. cloth, 10s. 6d.

FREDERICK SIMMS, M.B., M.R.C.P.

A WINTER IN PARIS: being a few Experiences and Observations of French Medical and Sanitary Matters. Fcap. 8vo. cloth, 4s.

E. B. SINCLAIR, M.D., F.K.Q.C.P., AND G. JOHNSTON, M.D., F.K.Q.C.P.

PRACTICAL MIDWIFERY: Comprising an Account of 13,748 Deliveries, which occurred in the Dublin Lying-in Hospital, during a period of Seven Years. 8vo. cloth, 10s.

J. L. SIORDET, M.B.LOND., M.R.C.P.

MENTONE IN ITS MEDICAL ASPECT. Foolscap 8vo. cloth, 2s. 6d.

ALFRED SMEE, M.R.C.S., F.R.S.

GENERAL DEBILITY AND DEFECTIVE NUTRITION; their Causes, Consequences, and Treatment. Second Edition. Fcap. 8vo. cloth, 3s. 6d.

WM. SMELLIE, M.D.

OBSTETRIC PLATES: being a Selection from the more Important and Practical Illustrations contained in the Original Work. With Anatomical and Practical Directions. 8vo. cloth, 5s.

HENRY SMITH, F.R.C.S.

I.

ON STRICTURE OF THE URETHRA. 8vo. cloth, 7s. 6d.

II.

HÆMORRHOIDS AND PROLAPSUS OF THE RECTUM: Their Pathology and Treatment, with especial reference to the use of Nitric Acid. Third Edition. Fcap. 8vo. cloth, 3s.

III.

THE SURGERY OF THE RECTUM. Lettsomian Lectures. Second Edition. Fcap. 8vo. 3s. 6d.

JOHN SMITH, M.D., F.R.C.S.EDIN.

HANDBOOK OF DENTAL ANATOMY AND SURGERY, FOR THE USE OF STUDENTS AND PRACTITIONERS. Fcap. 8vo. cloth, 3s. 6d.

J. BARKER SMITH.

PHARMACEUTICAL GUIDE TO THE FIRST AND SECOND EXAMINATIONS. Crown 8vo. cloth, 6s. 6d.

W. TYLER SMITH, M.D., F.R.C.P.

A MANUAL OF OBSTETRICS, THEORETICAL AND PRACTICAL. Illustrated with 186 Engravings. Fcap. 8vo. cloth, 12s. 6d.

JOHN SNOW, M.D.

ON CHLOROFORM AND OTHER ANÆSTHETICS: THEIR ACTION AND ADMINISTRATION. Edited, with a Memoir of the Author, by Benjamin W. Richardson, M.D. 8vo. cloth, 10s. 6d.

J. VOSE SOLOMON, F.R.C.S.

TENSION OF THE EYEBALL; GLAUCOMA: some Account of
the Operations practised in the 19th Century. 8vo. cloth, 4s.

STANHOPE TEMPLEMAN SPEER, M.D.

PATHOLOGICAL CHEMISTRY, IN ITS APPLICATION TO
THE PRACTICE OF MEDICINE. Translated from the French of MM. BECQUEREL
and RODIER. 8vo. cloth, reduced to 8s.

J. K. SPENDER, M.D.LOND.

A MANUAL OF THE PATHOLOGY AND TREATMENT
OF ULCERS AND CUTANEOUS DISEASES OF THE LOWER LIMBS,
8vo. cloth, 4s.

PETER SQUIRE.

I.

A COMPANION TO THE BRITISH PHARMACOPÆIA.
Seventh Edition. 8vo. cloth, 10s. 6d.   II.

THE PHARMACOPÆIAS OF THE LONDON HOSPITALS,
arranged in Groups for easy Reference and Comparison. Second Edition. 18mo.
cloth, 5s.

JOHN STEGGALL, M.D.

I.

A MEDICAL MANUAL FOR APOTHECARIES' HALL AND OTHER MEDICAL
BOARDS. Twelfth Edition. 12mo. cloth, 10s.

II.

A MANUAL FOR THE COLLEGE OF SURGEONS; intended for the Use
of Candidates for Examination and Practitioners. Second Edition. 12mo. cloth, 10s.

III.

FIRST LINES FOR CHEMISTS AND DRUGGISTS PREPARING FOR EX-
AMINATION AT THE PHARMACEUTICAL SOCIETY.   Third Edition.
18mo. cloth, 3s. 6d.

WM. STOWE, M.R.C.S.

A TOXICOLOGICAL CHART, exhibiting at one view the Symptoms,
Treatment, and Mode of Detecting the various Poisons, Mineral, Vegetable, and Animal.
To which are added, concise Directions for the Treatment of Suspended Animation.
Twelfth Edition. revised. On Sheet, 2s.; mounted on Roller, 5s.

FRANCIS SUTTON, F.C.S.

A SYSTEMATIC HANDBOOK OF VOLUMETRIC ANALYSIS;
or, the Quantitative Estimation of Chemical Substances by Measure. With Engravings.
Post 8vo. cloth, 7s. 6d.

W. P. SWAIN, F.R.C.S.

INJURIES AND DISEASES OF THE KNEE-JOINT, and
their Treatment by Amputation and Excision Contrasted. Jacksonian Prize Essay.
With 36 Engravings. 8vo. cloth, 9s.

J. G. SWAYNE, M.D.

OBSTETRIC APHORISMS FOR THE USE OF STUDENTS
COMMENCING MIDWIFERY PRACTICE. With Engravings on Wood. Fourth
Edition. Fcap. 8vo. cloth, 3s. 6d.

SIR ALEXANDER TAYLOR, M.D., F.R.S.E.

THE CLIMATE OF PAU; with a Description of the Watering Places of the Pyrenees, and of the Virtues of their respective Mineral Sources in Disease. Third Edition. Post 8vo. cloth, 7s.

ALFRED S. TAYLOR, M.D., F.R.C.P., F.R.S.

I.

THE PRINCIPLES AND PRACTICE OF MEDICAL JURISPRUDENCE. With 176 Wood Engravings. 8vo. cloth, 28s.

II.

A MANUAL OF MEDICAL JURISPRUDENCE. Eighth Edition. With Engravings. Fcap. 8vo. cloth, 12s. 6d.

III.

ON POISONS, in relation to MEDICAL JURISPRUDENCE AND MEDICINE. Second Edition. Fcap. 8vo. cloth, 12s. 6d.

THEOPHILUS THOMPSON, M.D., F.R.C.P., F.R.S.

CLINICAL LECTURES ON PULMONARY CONSUMPTION; with additional Chapters by E. Symes Thompson, M.D. With Plates. 8vo. cloth, 7s. 6d.

ROBERT THOMAS, M.D.

THE MODERN PRACTICE OF PHYSIC; exhibiting the Symptoms, Causes, Morbid Appearances, and Treatment of the Diseases of all Climates. Eleventh Edition. Revised by ALGERNON FRAMPTON, M.D. 2 vols. 8vo. cloth, 28s.

SIR HENRY THOMPSON, F.R.C.S.

I.

STRICTURE OF THE URETHRA AND URINARY FISTULÆ; their Pathology and Treatment. Jacksonian Prize Essay. With Plates. Third Edition. 8vo. cloth, 10s.

II.

THE DISEASES OF THE PROSTATE; their Pathology and Treatment. With Plates. Third Edition. 8vo. cloth, 10s.

III.

PRACTICAL LITHOTOMY AND LITHOTRITY; or, An Inquiry into the best Modes of removing Stone from the Bladder. With numerous Engravings, 8vo. cloth, 9s.

IV.

CLINICAL LECTURES ON DISEASES OF THE URINARY ORGANS. With Engravings. Second Edition. Crown 8vo. cloth, 5s.

J. C. THOROWGOOD, M.D.LOND.

NOTES ON ASTHMA; its Nature, Forms and Treatment. Crown 8vo. cloth, 4s.

J. L. W. THUDICHUM, M.D., M.R.C.P.

I.

A TREATISE ON THE PATHOLOGY OF THE URINE, Including a complete Guide to its Analysis. With Plates, 8vo. cloth, 14s.

II.

A TREATISE ON GALL STONES: their Chemistry, Pathology, and Treatment. With Coloured Plates. 8vo. cloth, 10s.

E. J. TILT, M.D., M.R.C.P.

I.

ON UTERINE AND OVARIAN INFLAMMATION, AND ON
THE PHYSIOLOGY AND DISEASES OF MENSTRUATION. Third Edition.
8vo. cloth, 12s.

II.

A HANDBOOK OF UTERINE THERAPEUTICS AND OF
DISEASES OF WOMEN.   Third Edition.   Post 8vo. cloth, 10s.

III.

THE CHANGE OF LIFE IN HEALTH AND DISEASE: a
Practical Treatise on the Nervous and other Affections incidental to Women at the Decline
of Life.   Second Edition.   8vo. cloth, 6s.

GODWIN W. TIMMS, M.D., M.R.C.P.

CONSUMPTION: its True Nature and Successful Treatment.   Re-issue,
enlarged.   Crown 8vo. cloth, 10s.

ROBERT B. TODD, M.D., F.R.S.

I.

CLINICAL LECTURES ON THE PRACTICE OF MEDICINE.
New Edition, in one Volume, Edited by DR. BEALE, 8vo. cloth, 18s.

II.

ON CERTAIN DISEASES OF THE URINARY ORGANS, AND
ON DROPSIES.   Fcap. 8vo. cloth, 6s.

JOHN TOMES, F.R.S.

A MANUAL OF DENTAL SURGERY. With 208 Engravings on
Wood.   Fcap. 8vo. cloth, 12s. 6d.

JAS. M. TURNBULL, M.D., M.R.C.P.

I.

AN INQUIRY INTO THE CURABILITY OF CONSUMPTION,
ITS PREVENTION, AND THE PROGRESS OF IMPROVEMENT IN THE
TREATMENT.   Third Edition.   8vo. cloth, 6s.

II.

A PRACTICAL TREATISE ON DISORDERS OF THE STOMACH
with FERMENTATION; and on the Causes and Treatment of Indigestion, &c.   8vo.
cloth, 6s.

R. V. TUSON, F.C.S.

A PHARMACOPŒIA; including the Outlines of Materia Medica
and Therapeutics, for the Use of Practitioners and Students of Veterinary Medicine.
Post 8vo. cloth, 7s.

ALEXR. TWEEDIE, M.D., F.R.C.P., F.R.S.

CONTINUED FEVERS: THEIR DISTINCTIVE CHARACTERS,
PATHOLOGY, AND TREATMENT.   With Coloured Plates.   8vo. cloth, 12s.

DR. UNDERWOOD.

TREATISE ON THE DISEASES OF CHILDREN. Tenth Edition,
with Additions and Corrections by HENRY DAVIES, M.D.   8vo. cloth, 15s.

VESTIGES OF THE NATURAL HISTORY OF CREATION.
Eleventh Edition.   Illustrated with 106 Engravings on Wood.   8vo. cloth, 7s. 6d.

J. L. C. SCHROEDER VAN DER KOLK.

THE PATHOLOGY AND THERAPEUTICS OF MENTAL
DISEASES. Translated by Mr. Rudall, F.R.C.S. 8vo. cloth, 7s. 6d.

MISS VEITCH.

HANDBOOK FOR NURSES FOR THE SICK. Crown 8vo.
cloth, 2s. 6d.

ROBERT WADE, F.R.C.S.

STRICTURE OF THE URETHRA, ITS COMPLICATIONS
AND EFFECTS; a Practical Treatise on the Nature and Treatment of those
Affections. Fourth Edition. 8vo. cloth, 7s. 6d.

ADOLPHE WAHLTUCH, M.D.

A DICTIONARY OF MATERIA MEDICA AND THERA-
PEUTICS. 8vo. cloth, 15s.

J. WEST WALKER, M.B.LOND.

ON DIPHTHERIA AND DIPHTHERITIC DISEASES. Fcap.
8vo. cloth, 3s.

CHAS. WALLER, M.D.

ELEMENTS OF PRACTICAL MIDWIFERY; or, Companion to
the Lying-in Room. Fourth Edition, with Plates. Fcap. cloth, 4s. 6d.

HAYNES WALTON, F.R.C.S.

SURGICAL DISEASES OF THE EYE. With Engravings on
Wood. Second Edition. 8vo. cloth, 14s.

E. J. WARING, M.D., M.R.C.P.LOND.

I.

A MANUAL OF PRACTICAL THERAPEUTICS. Second Edition,
Revised and Enlarged. Fcap. 8vo. cloth, 12s. 6d.

II.

THE TROPICAL RESIDENT AT HOME. Letters addressed to
Europeans returning from India and the Colonies on Subjects connected with their Health
and General Welfare. Crown 8vo. cloth, 5s.

A. T. H. WATERS, M.D., F.R.C.P.

I.

DISEASES OF THE CHEST. CONTRIBUTIONS TO THEIR
CLINICAL HISTORY, PATHOLOGY, AND TREATMENT. With Plates.
8vo. cloth, 12s. 6d.

II.

THE ANATOMY OF THE HUMAN LUNG. The Prize Essay
to which the Fothergillian Gold Medal was awarded by the Medical Society of London.
Post 8vo. cloth, 6s. 6d.

III.

RESEARCHES ON THE NATURE, PATHOLOGY, AND
TREATMENT OF EMPHYSEMA OF THE LUNGS, AND ITS RELA-
TIONS WITH OTHER DISEASES OF THE CHEST. With Engravings. 8vo.
cloth, 5s.

ALLAN WEBB, M.D., F.R.C.S.L.

THE SURGEON'S READY RULES FOR OPERATIONS IN
SURGERY. Royal 8vo. cloth, 10s. 6d.

J. SOELBERG WELLS.

I.

A TREATISE ON THE DISEASES OF THE EYE. With
Coloured Plates and Wood Engravings. Second Edition. 8vo. cloth, 24s.

II.

ON LONG, SHORT, AND WEAK SIGHT, and their Treatment by
the Scientific Use of Spectacles. Third Edition. With Plates. 8vo. cloth, 6s.

T. SPENCER WELLS, F.R.C.S.

SCALE OF MEDICINES FOR MERCHANT VESSELS.
With Observations on the Means of Preserving the Health of Seamen, &c. &c.
Seventh Thousand. Fcap. 8vo. cloth, 3s. 6d.

CHARLES WEST, M.D., F.R.C.P.

LECTURES ON THE DISEASES OF WOMEN. Third Edition.
8vo. cloth, 16s.

J. A. WHEELER.

HAND-BOOK OF ANATOMY FOR STUDENTS OF THE
FINE ARTS. With Engravings on Wood. Fcap. 8vo., 2s. 6d.

JAMES WHITEHEAD, M.D., M.R.C.P.

ON THE TRANSMISSION FROM PARENT TO OFFSPRING
OF SOME FORMS OF DISEASE, AND OF MORBID TAINTS AND
TENDENCIES. Second Edition. 8vo. cloth, 10s. 6d.

C. J. B. WILLIAMS, M.D., F.R.C.P., F.R.S.

PRINCIPLES OF MEDICINE: An Elementary View of the Causes,
Nature, Treatment, Diagnosis, and Prognosis, of Disease. With brief Remarks on
Hygienics, or the Preservation of Health. The Third Edition. 8vo. cloth, 15s.

FORBES WINSLOW, M.D., D.C.L.OXON.

OBSCURE DISEASES OF THE BRAIN AND MIND.
Fourth Edition. Carefully Revised. Post 8vo. cloth, 10s. 6d.

T. A. WISE, M.D., F.R.C.P.EDIN.

REVIEW OF THE HISTORY OF MEDICINE AMONG
ASIATIC NATIONS. Two Vols. 8vo. cloth, 16s.

ERASMUS WILSON, F.R.C.S., F.R.S.

I.

THE ANATOMIST'S VADE-MECUM: A SYSTEM OF HUMAN ANATOMY. With numerous Illustrations on Wood. Eighth Edition. Foolscap 8vo. cloth, 12s. 6d.

II.

ON DISEASES OF THE SKIN: A SYSTEM OF CUTANEOUS MEDICINE. Sixth Edition. 8vo. cloth, 18s.

THE SAME WORK; illustrated with finely executed Engravings on Steel, accurately coloured. 8vo. cloth, 36s.

III.

HEALTHY SKIN : A Treatise on the Management of the Skin and Hair in relation to Health. Seventh Edition. Foolscap 8vo. 2s. 6d.

IV.

PORTRAITS OF DISEASES OF THE SKIN. Folio. Fasciculi I. to XII., completing the Work. 20s. each. The Entire Work, half morocco, £13.

V.

THE STUDENT'S BOOK OF CUTANEOUS MEDICINE AND DISEASES OF THE SKIN. Post 8vo. cloth, 8s. 6d.

VI.

LECTURES ON EKZEMA AND EKZEMATOUS AFFEC- TIONS; with an Introduction on the General Pathology of the Skin, and an Appendix of Essays and Cases. 8vo. cloth, 10s. 6d.

VII.

ON SYPHILIS, CONSTITUTIONAL AND HEREDITARY; AND ON SYPHILITIC ERUPTIONS. With Four Coloured Plates. 8vo. cloth, 16s.

VIII.

A THREE WEEKS' SCAMPER THROUGH THE SPAS OF GERMANY AND BELGIUM, with an Appendix on the Nature and Uses of Mineral Waters. Post 8vo. cloth, 6s. 6d.

IX.

THE EASTERN OR TURKISH BATH: its History, Revival in Britain, and Application to the Purposes of Health. Foolscap 8vo., 2s.

G. C. WITTSTEIN.

PRACTICAL PHARMACEUTICAL CHEMISTRY: An Explanation of Chemical and Pharmaceutical Processes, with the Methods of Testing the Purity of the Preparations, deduced from Original Experiments. Translated from the Second German Edition, by STEPHEN DARBY. 18mo. cloth, 6s.

HENRY G. WRIGHT, M.D., M.R.C.P.

I.

UTERINE DISORDERS: their Constitutional Influence and Treatment. 8vo. cloth, 7s. 6d.

II.

HEADACHES; their Causes and their Cure. Fourth Edition. Fcap. 8vo. 2s. 6d.

CHURCHILL'S SERIES OF MANUALS.

Fcap. 8vo. cloth, 12s. 6d. each.

"We here give Mr. Churchill public thanks for the positive benefit conferred on the Medical Profession, by the series of beautiful and cheap Manuals which bear his imprint."— *British and Foreign Medical Review.*

AGGREGATE SALE, 160,000 COPIES.

ANATOMY. With numerous Engravings. Eighth Edition. By ERASMUS WILSON, F.R.C.S., F.R.S.

BOTANY. With numerous Engravings. Second Edition. By ROBERT BENTLEY, F.L.S., Professor of Botany, King's College, and to the Pharmaceutical Society.

CHEMISTRY. With numerous Engravings. Tenth Edition, 14s. By GEORGE FOWNES. F.R.S., H. BENCE JONES, M.D., F.R.S., and HENRY WATTS, B.A., F.R.S.

DENTAL SURGERY. With numerous Engravings. By JOHN TOMES, F.R.S.

EYE, DISEASES OF. With coloured Plates and Engravings on Wood. By C. MACNAMARA.

MATERIA MEDICA. With numerous Engravings. Fifth Edition. By J. FORBES ROYLE, M.D., F.R.S., and F. W. HEADLAND, M.D., F.R.C.P.

MEDICAL JURISPRUDENCE. With numerous Engravings. Eighth Edition. By ALFRED SWAINE TAYLOR, M.D., F.R.S.

PRACTICE OF MEDICINE. Second Edition. By G. HILARO BARLOW, M.D., M.A.

The MICROSCOPE and its REVELATIONS. With numerous Plates and Engravings. Fourth Edition. By W. B. CARPENTER, M.D., F.R.S.

NATURAL PHILOSOPHY. With numerous Engravings. Sixth Edition. By CHARLES BROOKE, M.B., M.A., F.R.S. *Based on the Work of the late Dr. Golding Bird.*

OBSTETRICS. With numerous Engravings. By W. TYLER SMITH, M.D., F.R.C.P.

OPHTHALMIC MEDICINE and SURGERY. With coloured Plates and Engravings on Wood. Third Edition. By T. WHARTON JONES, F.R.C.S., F.R.S.

PATHOLOGICAL ANATOMY. With numerous Engravings. By C. HANDFIELD JONES, M.B., F.R.S., and E. H. SIEVEKING, M.D., F.R.C.P.

PHYSIOLOGY. With numerous Engravings. Fourth Edition. By WILLIAM B. CARPENTER, M.D., F.R.S.

POISONS. Second Edition. By ALFRED SWAINE TAYLOR, M.D., F.R.S.

PRACTICAL ANATOMY. With numerous Engravings. Second Edition. By CHRISTOPHER HEATH, F.R.C.S.

THERAPEUTICS. Second Edition. By E. J. Waring, M.D., M.R.C.P.

www.ingramcontent.com/pod-product-compliance
Lightning Source LLC
Chambersburg PA
CBHW021106270326
41929CB00009B/755